13th International Conference on 3D Dosimetry (IC3DDose 2024)

Journal of Physics: Conference Series Volume 2799

Aarhus, Denmark
17-19 June 2024

ISBN: 979-8-3313-0211-5
ISSN: 1742-6588

Printed from e-media with permission by:

Curran Associates, Inc.
57 Morehouse Lane
Red Hook, NY 12571

Some format issues inherent in the e-media version may also appear in this print version.

Institute of Physics
Dirac House, Temple Back
Bristol BS1 6BE UK

Phone: 44 1 17 929 7481
Fax: 44 1 17 920 0979

techtracking@iop.org

Additional copies of this publication are available from:

Curran Associates, Inc.
57 Morehouse Lane
Red Hook, NY 12571 USA
Phone: 845-758-0400
Fax: 845-758-2633
Email: curran@proceedings.com
Web: www.proceedings.com

TABLE OF CONTENTS

Journal of Physics: Conference Series **2799** (2024) 011001 doi:10.1088/1742-6596/2799/1/011001

IC3DDose 2024 – 13th International Conference on 3D Dosimetry

Preface

The 13th International Conference on 3D Dosimetry was held in Aarhus, Denmark, on 17-19 June 2024.

The IC3DDose conference serves as a meeting point between scientists developing new dosimeter systems and readout methods and researchers with a clinical background and motivation. Participants included both new and more experienced researchers in the field as well as company representatives.

A total of 66 participants from 17 countries met for three days at the Aarhus University campus in order to present and discuss the latest developments in advanced radiation dosimetry, predominately with applications towards three-dimensional dose measurements. One opening address, 12 invited talks, 20 contributed talks, and 10 poster presentations were given at the conference. In addition, the optimal ways to bring the scientific results into the radiotherapy clinics, for the benefit of patients, were discussed at a panel debate. All activities took place in a single track, and the participants had good possibilities to interact during designated poster sessions, meet-the-speakers-of-the-day sessions, and breaks. The setting provided a homely campus atmosphere that inspired the participants to engage in numerous scientific discussions.

The conference also included a welcome reception at Danish Centre for Particle Therapy (DCPT), where conference participants were invited on a tour behind the scenes at the proton facility.

The 13th edition of the IC3Ddose conference contributed to the continuous identification of new research areas of relevance for the overarching theme of three-dimensional dose measurements. It brought together scientists from the more established chemically based 3D dosimeters with scientists developing approaches based on physical processes, such as scintillation, optically stimulated luminescence, Cherenkov radiation, and charge-based technologies. This mix appeared to be a source of mutual inspiration and cross-fertilization of both scientific ideas and techniques. Several of the developments presented at the conference showed potential to become useful for the treatment verification of future cancer patients.

The associated 21 papers presented in these proceedings add to an already extensive collection of 118 review and 612 proffered papers in 10 previous volumes of the Journal of Physics Conference Series from the past conferences.

International Scientific Committee

Louis Archambault, Université Laval, Canada
Clive Baldock, Western Sydney University, Australia
Peter Balling, Aarhus University, Denmark
Luc Beaulieu, Université Laval, Canada
Sam Beddar, MD Anderson, USA

Journal of Physics: Conference Series **2799** (2024) 011001 doi:10.1088/1742-6596/2799/1/011001

Sven Bäck, Lund University, Sweden
Sofie Ceberg, Lund University, Sweden
Yves De Deene, Southwestern Sydney Local Health District and Western Sydney University, Australia
Simon Doran, Institute of Cancer Research, UK
Geoffrey Ibbott, MD Anderson, USA
Andrew Jirasek, University of British Columbia, Okanagan, Canada
Marek Kozicki, Lodz University of Technology, Poland
Boyd McCurdy, University of Manitoba, Canada
Mark Oldham, Duke University, USA
John Schreiner, Queen's University, Canada

Local Organizing Committee

Claus Andersen, Danish Technical University, Denmark
Christina Ankjærgaard, Danish Technical University, Denmark
Peter Balling, Aarhus University, Denmark
Niels Bassler, Danish Centre for Particle Therapy (DCPT) and Aarhus University, Denmark
Sofie Ceberg, Lund University, Sweden
Jacob Johansen, Danish Centre for Particle Therapy (DCPT) and Aarhus University, Denmark
Brian Julsgaard, Aarhus University, Denmark
Ludvig Muren, Danish Centre for Particle Therapy (DCPT) and Aarhus University, Denmark
Jørgen Petersen, Danish Centre for Particle Therapy (DCPT) and Aarhus University, Denmark
Per Poulsen, Danish Centre for Particle Therapy (DCPT) and Aarhus University, Denmark
Rosana Turtos, Aarhus University, Denmark

Editors of the proceedings

Jacob Johansen, Danish Centre for Particle Therapy (DCPT) and Aarhus University, Denmark
Brian Julsgaard, Aarhus University, Denmark

Journal of Physics: Conference Series **2799** (2024) 011002 doi:10.1088/1742-6596/2799/1/011002

Peer review statement

All papers published in this volume have been reviewed through processes administered by the Editors. Reviews were conducted by expert referees to the professional and scientific standards expected of a proceedings journal published by IOP Publishing.

Contributors, who wished to present at the conference (oral or poster), had to submit an extended abstract (max. four pages) via a "Call for Papers" in the Morressier Proceedings Management System, which resulted in 33 submissions. These were all found acceptable for presentation by the international scientific committee of the conference.

The contributors had the option to publish their extended abstracts in the conference proceedings, and 21 out of the 33 contributors chose this option. Invited speakers were also given an opportunity to publish an extended abstract in the conference proceedings, and one speaker chose this option. In total, 22 manuscripts were thus intended for publication and sent for review.

Type of peer review: Single Anonymous
Conference submission management system: Morressier
Number of submissions received: 34
Number of submissions sent for review: 22
Number of submissions accepted: 21
Acceptance Rate (Submissions Accepted / Submissions intended for publication × 100): 95
Average number of reviews per paper: 4
Total number of reviewers involved: 18

Contact person for queries:
Name: Brian Julsgaard
E-mail: brianj@phys.au.dk
Affiliation: Aarhus University

Journal of Physics: Conference Series **2799** (2024) 012001 doi:10.1088/1742-6596/2799/1/012001

Multiphase 3D dosimeters for radiotherapy

Marek Kozicki[1,2*], Piotr Maras[3], Malwina Jaszczak-Kuligowska[1]

[1] Department of Mechanical Engineering, Informatics and Chemistry of Polymer Materials, Faculty of Materials Technologies and Textile Design, Lodz University of Technology, Lodz, Poland
[2]GeVero Co., Lodz, Poland
[3]Department of Radiotherapy Planning, Copernicus Hospital, Lodz, Poland

*E-mail: marek.kozicki@p.lodz.pl; gevero@polygevero.com

Abstract. 3D dosimeters are mainly manufactured as single-phase chemical compositions in a container. In most cases, they imitate a single soft tissue, such as muscle or the brain. Research on dosimeters imitating lungs and bones is very rare. In turn, in this work we report the first experiments aimed at producing multiphase 3D dosimeters imitating different tissues in one container.

1. Introduction

Rapidly developing patient irradiation techniques in radiotherapy require appropriate 3D dosimetric materials. The dosimeters fall into three categories: water equivalent (imitating e.g., muscle, brain, internal organs), low density lung equivalent and high-density bone equivalent materials [1-3]. Such dosimetry has been studied for about forty years and includes various 3D/4D dosimeters: Fricke gels, radiochromic gels (including deformable ones), plastic compositions, radiofluorogenic gels and polymer gel dosimeters [1,2]. These dosimeters are manufactured as single-phase systems, which means that one container holds only one composition intended to imitate water equivalent tissues, lungs or bones [1-3]. However, the human body is not a single-phase system and has tissues of different densities attaching each other, which translates into specific absorption of ionizing radiation. Therefore, we started researching complex, combined 3D dosimeters that mimic different tissues in one container. The first proposal was based on a two-phase radiochromic dosimeter, one transparent phase imitating water equivalent tissues and the other imitating lungs that attach to each other forming an interphase [4]. In turn, in this work we describe a two- and three-phase dosimeter imitating two or three tissues: lungs, e.g. muscles and bones, in one container based on a 3D polymer gel dosimeter.

There are several challenges associated with such dosimeters. First, each phase should be designed to mimic a specific tissue. The basic dose-response characteristics of each part must then be examined separately. Finally, methods for manufacturing multiphase dosimeters should be investigated. Ultimately, such a multiphase dosimeter can be tested for applications in radiotherapy, for example to verify the dose distribution (planned, using a treatment planning system) at the interphases and in each phase of the dosimeter. Dedicated algorithms should also be developed to calculate dose distributions in each phase of the dosimeter. This work introduces readers to the subject of multiphase dosimeters and issues related to them.

2. Components of a multiphase dosimeter

Multiphase dosimeters can be manufactured based on any 3D polymer gel dosimeter imitating muscles (brain; water-equivalent). However, this requires modifications to obtain compositions imitating lungs or bones. Our first attempts were based on water-equivalent PAGAT2-Pluronic F-127 dosimeter imitating muscles or brain (Figure 1A), of a CT signal of about 28 HU, density=1.055 g/cm^3, and Z_{eff} (effective atomic number) of 7.26 [5]. The research was carried out in two ways: (i) towards a lung imitating dosimeter by foaming the PAGAT2-Pluronic F-127 solution (Figure 1B) and (ii) towards a bone imitating dosimeter by adding the main bone component: hydroxyapatite (Figure 1C) [6]. The conclusions regarding the lung-imitating part are as follows. A lung-imitating dosimeter (exhale) can be obtained by foaming the solution to approximately double its volume. The foam should be prepared in an oxygen-free environment, e.g. in a glove box. When measuring such foam using iterative cone beam computed tomography (iCBCT, TrueBeam, Varian, USA), the values oscillate around -500 HU (Z_{eff} does not change). As concerns the bone-imitating part, adding water-insoluble inorganic hydroxyapatite microparticles to the 3D polymer gel dosimeter solution may have a counterproductive effect; rapid sedimentation in the case of gelatine-based dosimeters. This is counteracted with a dosimetric composition based on Pluronic F-127, because this copolymer sets very quickly (unlike gelatine) and thus binds the hydroxyapatite particles in the dosimetric substance. Using different hydroxyapatites with different concentrations, it is possible to obtain a bone-like dosimeter imitating hard, compact bones or trabecular bones. For instance, 18% hydroxyapatite with a density of 0.59 g/cm^3 gives a bone-like structure imitating trabecular bone with a density of 1.16 g/cm^3, Z_{eff} of 10.47, and gives a CT signal of ~260 HU, which is similar to the trabecular bone signal of ~239 HU.

A B C

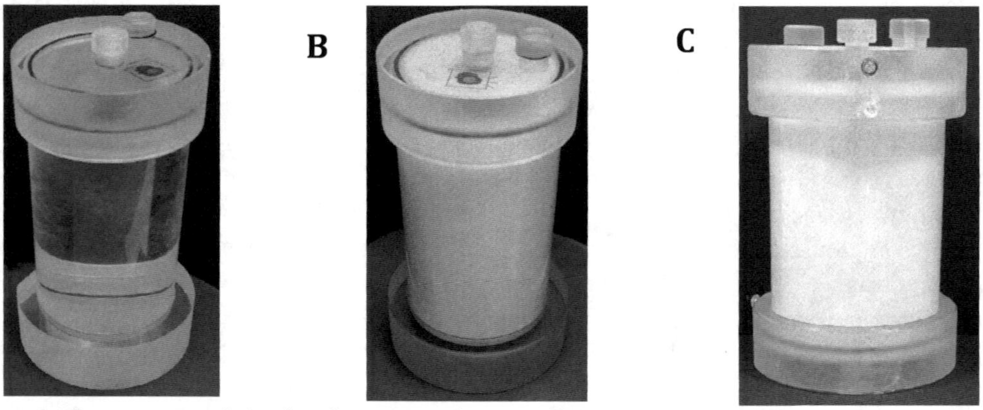

Figure 1. Three types of single-phase 3D polymer gel dosimeters in GeVero Co. containers: transparent imitating muscles (A), white imitating lungs (foam) (B), and white imitating trabecular bone (C). All dosimeters were made based on the PAGAT2-Pluronic F-127 dosimeter. The bone imitating dosimeter contains hydroxyapatite (18% (w/w); H121722, bulk density: 0.59 g/cm^3, Merz North America, USA).

3. Manufacturing possibilities: multiphase dosimeters

It is possible to manufacture two- or three-phase dosimeters using previously elaborated compositions imitating muscles, lungs and bones. The rapid solidification of the PAGAT2-Pluronic F-127 solution thanks to the specific phase transition of the Pluronic F-127 copolymer

[5] allows the shaping of parts (volumes) corresponding to various tissues. Examples of two- and three-phase dosimeters are presented in Figure 2-3. In Figure 2A, another two-phase dosimeter is presented, but it consists of two parts imitating the lungs during exhalation and an intermediate part imitating the muscles; these parts are diagonal to each other. Figure 2B provides a cross-sectional image of the iCBCT to visualize the internal parts of this combined dosimeter. Figure 3A shows a three-phase dosimeter; from top to bottom: part imitating lung, part imitating muscle, part imitating bone and part imitating muscle. Figure 3B corresponds to a cross-section illustrating the internal part of the dosimeter with different phases.

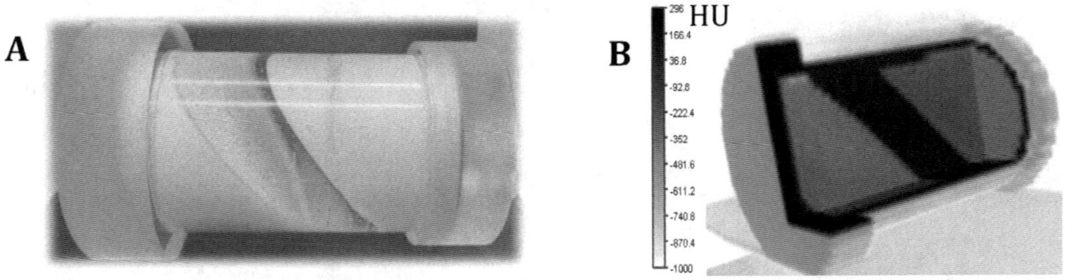

Figure 2. Two-phase dosimeter imitating lungs (white area) and muscles (transparent area). A: Photo of the dosimeter in a GeVero Co. container. B: 3D visualisation of the dosimeter using polyGeVero software (GeVero Co., Poland) after iCBCT scanning (TrueBeam, Varian, USA). Both phases are based on the PAGAT2-Pluronic F-127 dosimeter.

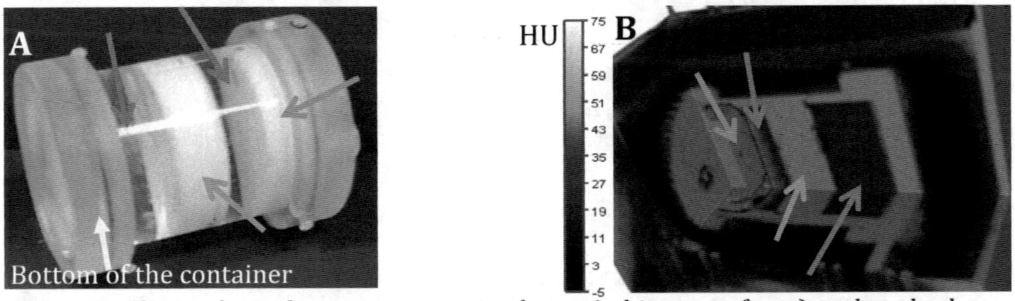

Figure 3. Three-phase dosimeter imitating lungs (white area, foam), trabecular bone (white region) and muscles (transparent part). A: Photo of the dosimeter in a GeVero Co. container. B: 3D visualisation of the dosimeter using polyGeVero software (GeVero Co., Poland) after CT scanning (Siemens Sensation Open, Germany). The green arrows point to the lung part, the red arrows point to the muscle part, and the blue arrows point to the bone part. Bone imitating part contains 18% hydroxyapatite (Merz North America, USA).

4. Data processing

Multiphase dosimeters can be read in 3D using CT that requires image data processing. For example, CT data processing requires, among other things, background subtraction to enhance the signal for irradiated areas [7,8]. Converting multiphase dosimeters into a three-dimensional dose matrix requires dedicated algorithms to delineate areas/volumes that mimic different tissues and apply specific calibrations to these areas. The polyGeVero-CT software package has been adapted for data processing that requires the calculation of a 3D dose distribution for each part of the combined dosimeters using calibration equations specific to these parts of the dosimeter, imitating e.g. bones, lungs and muscles. The view of dedicated functionalities in polyGeVero-CT is presented in Figure 4. Most importantly, they contain tools for determining the volumes appropriate for dosimeter parts and using previously obtained calibrations for these parts from the calibration and contour database, dose distributions can be calculated. After these operations, the entire 3D dose matrix can be exported for further processing in the polyGeVero software package (e.g. for comparison with TPS data; GeVero Co.).

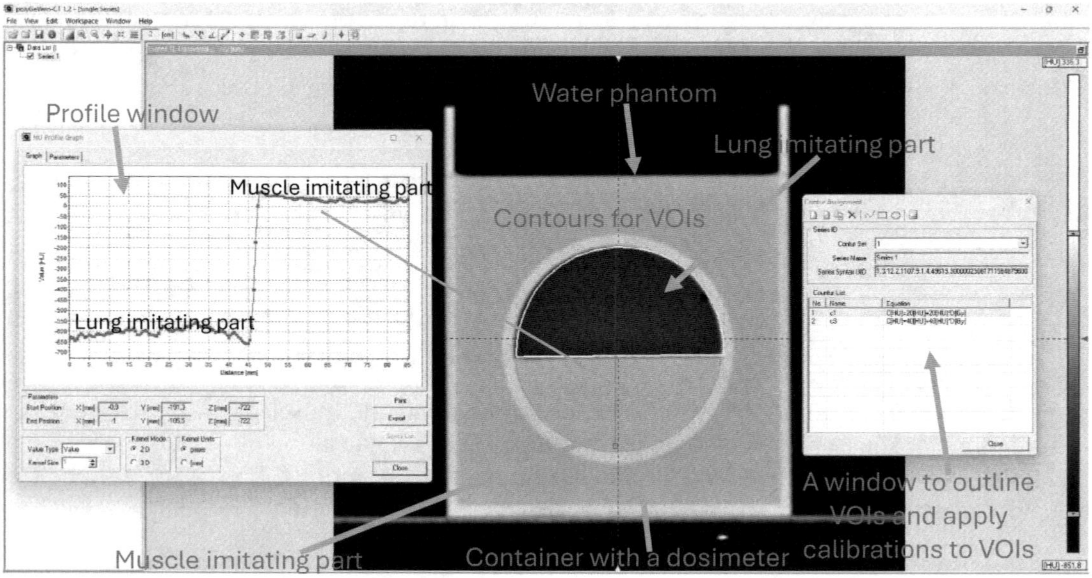

Figure 4. One of the three workspaces of the polyGeVero-CT software package (GeVero Co., Poland) during processing a DICOM series (iterative cone beam CT, iCBCT, TrueBeam) for a multiphase 3D polymer gel dosimeter imitating lungs and muscles in one container.

5. Conclusions

Multiphase dosimeters can be prepared to better suit the anatomical structure of the human body. Current research focuses on the dose response characteristics of such dosimeters and their application in radiotherapy. It seems that such compositions may also be measured using magnetic resonance imaging, which has not yet been tested for such complex dosimeters.

Acknowledgement: The authors kindly thank Merz North America, USA, for providing the hydroxyapatite component for this study.

References

[1] Baldock C *et al* 2010 *Phys. Med. Biol.* **55** R1
[2] De Deene Y 2022 *Gels* **8** 599
[3] Kumahara N *et al* 2023 *Radiol. Phys. Technol.* **16** 227
[4] Kozicki M *et al* 2023 *Adv. Mater. Technol.* **8** 2201023
[5] Jaszczak M *et al* 2018 *Phys. Med. Biol.* **63** 175010
[6] Jaszczak-Kuligowska M *et al* 2024, IC3DDose 2024, IOP *J. Phys.: Conf. Ser.* submitted.
[7] Maras P and Kozicki M 2022 *Materials* **15** 6807
[8] Johnston H *et al* 2012 *Phys. Med. Biol.* **57** 3155

Journal of Physics: Conference Series **2799** (2024) 012002 doi:10.1088/1742-6596/2799/1/012002

Reusable optically stimulated luminescence dosimetry films for 2D dose verification of proton therapy

S. J. G. Clausen[1,2], M. L. Jensen[2], C. L. Nielsen[2,3], M. F. Jensen[1], L. Muren[1,4], C. S. Søndergaard[1], P. S. Skyt[1], L. B. Stick[1], E. Yates[5], S. H. Vindbæk[1,4], P. Balling[2,3]

[1] Danish Centre for Particle Therapy, Aarhus University Hospital, Aarhus, Denmark
[2] Department of Physics and Astronomy, Aarhus University, Aarhus, Denmark
[3] Interdisciplinary Nanoscience Center, Aarhus University, Aarhus, Denmark
[4] Department of Clinical Medicine, Aarhus University, Aarhus, Denmark
[5] Department of Medical Physics, Aarhus University Hospital, Aarhus, Denmark

Email: clausen@au.dk

Abstract Accurate surface-dose measurements in proton therapy are challenging and often of poor spatial resolution for most dosimeter types. However, optically stimulated luminescence (OSL) based 2D dosimeters could provide the required spatial dose resolution. The aim of this study was to investigate the dosimetric precision and energy dependence of an in-house made reusable dosimeter composed of a silicone-film containing OSL-active nanoparticles. The dosimeter was irradiated with a clinical proton therapy field and the readout dose was compared to the results from a commercial dosimeter. The pixelwise noise-to-signal ratio for the OSL dosimeter remained below 2% for doses above 1 Gy, and the energy dependence was negligible in the investigated energy range.

1. Introduction

In radiotherapy, treatment planning systems (TPSs) are used to perform dose calculations for patients. Such dose calculations must be validated through measurements to ensure the doses delivered to patients are correct, and to highlight potential uncertainties in the treatment plan. In particular, air-tissue interfaces such as the skin and internal air cavities can lead to considerable uncertainties in dose calculations [1]. Verifying the dose is crucial since both over- and under-dosages can have detrimental consequences. Patient-specific dose verification is typically conducted using a 2D ionization chamber array for measuring specific treatment plans, but such systems can have a limited spatial resolution. Furthermore, to obtain charge particle equilibrium, the chamber must be of a certain depth, and they can therefore only measure delivered dose at depths exceeding several millimetres, which is a problem when the goal is to verify calculated skin and surface doses.

Skin dose can be measured using thermoluminescence point dosimeters [2] and radiochromic films [3]. However, radiochromic films, currently the only 2D option, are not reusable nor mechanically sufficiently flexible, and they require batch-to-batch calibration.

Silicone films containing optically stimulated luminescence (OSL)-active nanoparticles could provide a new and reusable solution [4]. The silicone films only need an initial calibration and are flexible. This allows the dosimeter to conform to a surface and mimic the skin or the surface of an internal organ. Furthermore, they are applicable for both proton- and photon-based radio therapy.

In the past, OSL point dosimeters have been used for 2D dose mapping. Recently, a successful approximation of skin-dose measurements using OSL point detectors in vivo was demonstrated, showing the potential for future clinical use [5]. Using point dosimeters has however been shown to be time-consuming and associated with poor spatial resolution and complicated procedures [5-6]. However, by producing OSL area detectors as films, OSL systems have shown potential for evaluating doses for 2D mapping relatively quickly and simply, with much better spatial resolution [7-8], and a large-area (20 cm x 18 cm) 2D OSL film has previously been developed showing the potential for a

2D dosimeter based on OSL [9]. Furthermore, an in-house silicone dosimeter containing $LiBaF_3$ and LiF:Cu nanoparticles has shown promising qualities as a candidate for OSL dosimetry [10,11]. Moreover, it has been suggested that OSL detectors hold promise for future applications in 3D dose verifications: Lately, OSL 3D measurements have been performed in a YSO:Ce crystal [12], while investigations into developing a tissue-equivalent 3D OSL dosimeter are still ongoing [12]. However, clinical demonstrations of 2D and 3D dosimetry have to the best of our knowledge not yet been reported.

The aim of this study was to explore the dosimetric properties of a 2D in-house made reusable OSL dosimeter for dose verification of proton therapy, by investigating the energy dependence, resolution and dosimetric precision of the dosimeter.

2. Material and methods

2.1. Samples

A 2D silicone film (100 x 100 x 1.6 mm^3) containing OSL-active nanoparticles of $LiBaF_3$ and LiF:Cu was used for all measurements. The $LiBaF_3$ and LiF:Cu nanoparticles were synthesized with a custom-built flow reactor using a one-step cold precipitation flow synthesis [10,11]. To produce the dosimeter, the nanoparticles (8% LiF:Cu and 0.8% $LiBaF_3$ by weight) were embedded in Dow Corning Sylgard 184 using a base to curing agent ratio of 5:1, and the mixture was poured into a square mould with an airtight lid to minimize air bubbles in the dosimeter and was left to cure for three days.

2.2 Irradiations

The film dosimeter was placed on top of an IBA Dosimetry MatriXX PT chamber array to compare the delivered treatment field doses and was covered by 1.7 cm of a water-equivalent material. The OSL dosimeter was irradiated with a nonuniform clinical pencil beam scanning proton field (Varian ProBeam) spanning 10 x 10 cm^2 and consisting of 547 spots, with 28 energy layers in the range 72-176 MeV, with water equivalent doses varying from 0 to 1 Gy. The source-to-surface distance was 98.6 cm, and a 3 cm range shifter was applied.

To investigate the energy dependence, the dosimeter was irradiated with varying entrance mono-energetic proton fields ranging from 100-220 MeV. All measurements were performed with 5 cm of a water-equivalent material under and 3 cm on top of the dosimeter, hence effectively irradiating the dosimeter in the energy range ~80-200 MeV. To minimize potential radiative contributions to the dosimeter from the water-equivalent material, the material was switched between each measurement and irradiation.

2.3 Readout and data analysis

The OSL signal was stimulated using 451 nm blue LED illumination and captured using a UV objective which images the film onto a charged-coupled device (CCD) camera with a spatial resolution on the film of 1.16 x 1.16 mm^2 (defined by hardware binning and the objective magnification). A set of optical filters was applied to block the stimulating blue light but allowing the ultraviolet luminescence of the OSL signal to pass [12]. Once read out, the dosimeter was optically bleached using the blue LEDs to ensure that the dosimeter was reset before background measurements were taken and further irradiations and readouts could be made. After each measurement, the background was subtracted from the measurements, single bright pixels from cosmic-ray events removed, and an optical roll-off correction was performed [12].

By comparing the measurements from the energy-dependence experiments to the calculated water-equivalent doses at the specified depth, a dose-response in counts per Gy for the whole dosimeter was found, and the energy dependence was derived. It was noted that for the lowest energy, the calculated dose varied considerably along the depth (thickness) of the dosimeter, which contributed to the final uncertainty in the counts per Gy for especially the lowest energies. Using the dose-response, the

Journal of Physics: Conference Series **2799** (2024) 012002 doi:10.1088/1742-6596/2799/1/012002

processed signal from the clinical irradiation of the dosimeter was converted to water-equivalent dose and software binned into 7.6 x 7.6 mm^2 pixels to compare it to the 2D ionization chamber array measurement of the same irradiation field.

3. Results

Comparing the binned readout dose from the OSL dosimeter and the 2D ionization chamber array measurement showed good dosimetric agreement (Fig. 1). The unbinned dose distribution of the OSL dosimeter had a significantly higher spatial resolution than the chamber array, thus providing more detail on the dose variations, and even at this higher resolution, its pixel-wise noise-to-signal ratio remained below 2% for doses above 1 Gy (Fig. 1). The unbinned, total dose response for the OSL dosimeter was $(1.97 \pm 0.02) \cdot 10^7$ counts per Gy and its energy dependence was close to negligible in the investigated energy range (Fig. 2).

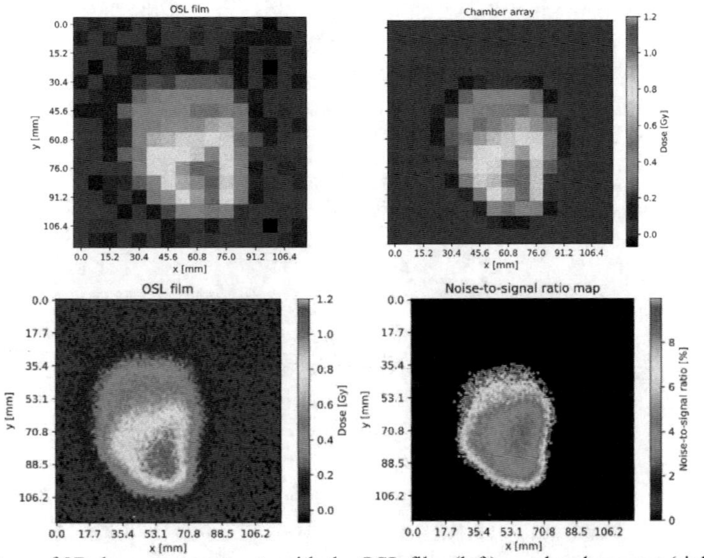

Figure 1: Upper: Comparison of 2D dose measurements with the OSL film (left) vs. chamber array (right). The irradiations were performed at a water equivalent depth of 1.7 cm. The OSL dosimeter data was binned to match the resolution of the ionisation chamber array. Lower: Left: The dose distribution measured by the OSL film without software binning. Right: The noise-to-signal ratio for the OSL dosimeter in percent. The image only displays the noise-to-signal ratio for pixels that received more than 10% of the maximum dose.

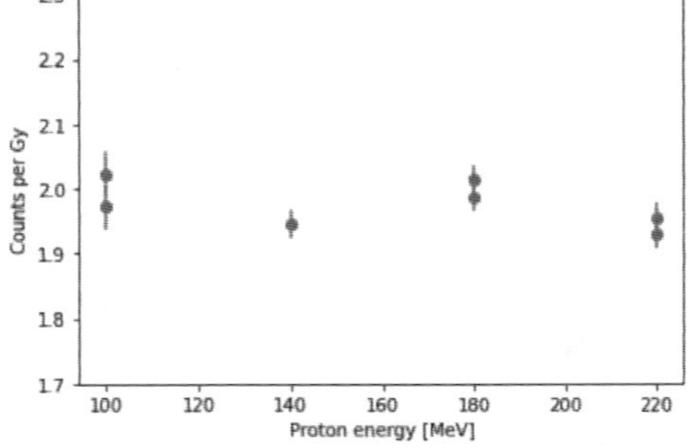

Figure 2: The measured dose response per Gy for the whole dosimeter for four different proton entrance energies. All irradiations were performed at a water equivalent depth of 3 cm. Some of the measurements have been redone for statistics. The uncertainties arise from both the calculated dose in the TPS (due to algorithm selection, beam model, etc.) and the uncertainty associated with the actual dose delivered to the film compared to the calculated dose (machine output variability and measurement setup not necessarily matching the calculated geometry perfectly).

Journal of Physics: Conference Series **2799** (2024) 012002 doi:10.1088/1742-6596/2799/1/012002

4. Discussion

The dosimetric precision of the OSL dosimeter was investigated by irradiating the OSL dosimeter and ionization chamber array with identical proton fields and comparing the measurements. The energy dependence was examined by irradiating the dosimeter with varying energies, and for each energy normalized to the calculated, delivered dose. The OSL dosimeter showed good dosimetric agreement with the conventional ionization chamber array, and the energy dependence appeared to be negligible in the investigated energy range.

For lower energies where the LET would increase, the energy dependence is unknown, and further studies would be needed.

The OSL dosimeter has been calibrated to the same absolute dosimetry as the chamber array, and hence its ability to measure absolute dose cannot be validated, but the results still indicated that the 2D OSL dosimeter could be a promising material for future dosimetric verifications, since even outside of the range investigated for energy dependence, the dosimeter had a clear resemblance with the chamber array. The resemblance also indicated a linear dose response in the clinical dose range, in low-LET areas, which harmonizes with previous studies [5,9,13].

It should be noted that the noise-to-signal ratio for the OSL dosimeter is still higher than for modern commercial 2D dosimeters used in quality assurance when the highest spatial resolution is used. However, the noise-to-signal ratio decreases if the signal is binned to larger pixels or if the delivered dose is increased.

While previous studies have produced OSL dosimeters covering larger areas [9], the flexibility of our dosimeter could enable 2.5D dosimetric measurements, by curving the dosimeter around the surface of phantoms. In the future, the 2D OSL dosimeter will be further characterized and used for various surface-dose measurements, e.g. skin dose and the surface of internal air cavities in a head-and-neck phantom, in both photon and proton therapy. However, before taking advantage of the flexibility of the dosimeter, its angular dependence of proton beams needs to be investigated. When fully characterized and tested, the plan is to apply the dosimeter also for *in-vivo* investigations.

In conclusion, it has been shown that the OSL film is a promising 2D high resolution dosimeter with high dosimetric accuracy in clinical dose and energy ranges.

5. References

[1] Asher D et al. *2019 Med Dosim.* ***44*** *405-408*
[2] Farhood B et al. *2018 J Cancer Res Ther.* ***14*** *1214-1219*
[3] Magnier F et al. *2018 Cardiovasc Intervent Radiol.* ***41*** *762*
[4] Nyemann J S et al *2023 J. Phys.: Conf. Ser.* ***2630*** *012022*
[5] Ratheesh K E et al *2024 Biomed. Phys. Eng. Express* ***10*** *045025*
[6] Gasparian P B R et al *2022 Phys. Med. Biol.* ***67*** *19*
[7] Jahn A et al *2014 Radiat. Meas.* ***56*** *324–7*
[8] Martins M S et al *2020 Braz. J. Radiat. Sci.* ***8*** *2*
[9] Wouter C et al. 2017, *Physics in Medicine and Biology* ***62*** *21*
[10] Nielsen C L et al. *2023 Journal of Luminescence* ***263*** *120024*
[11] Nielsen C L et al. *2022 Nano Lett.* ***22*** *1566-1572*
[12] Jensen M L et al. *2024 Med Phys.* ***51*** *2200-2209*
[13] Ahmed M F et. al *2017 Radiat. Meas.* ***106*** *315-320*

Journal of Physics: Conference Series **2799** (2024) 012003 doi:10.1088/1742-6596/2799/1/012003

Effects of signal source to camera distance in Cherenkov dosimetry using polarized imaging

Audran Poher[1,2*], Gérémy Michaud[1,2], Louis Archambault[1,2] and Luc Beaulieu[1,2]

[1] Service de physique médicale et de radioprotection, Centre Intégré de Cancérologie, CHU de Québec – Université Laval et Centre de recherche du CHU de Québec, Québec, Québec, Canada
[2] Département de physique, de génie physique et d'optique et Centre de recherche sur le cancer, Université Laval, Québec Québec G1V 0A6, Canada

*E-mail: audran.poher.1@ulaval.ca

Abstract. As Cherenkov emissions are related to dose, they can be used as signal for dosimetric purposes using polarized imaging. However, when using this method, angular corrections must be applied because of Cherenkov anisotropy. Several Cherenkov light source-to-camera distances were tested to characterize its impact on the deviations from reference values of the calculated dose distributions. Average relative differences ranging from -1.6% to -11% between Cherenkov-based results and scintillation results were extracted. It is observed that an increase in Cherenkov light source-to-camera distance decreases the difference between the Cherenkov optical signal and the dose. This reduces the importance of angular corrections at higher distances (4 m and beyond).

1. Introduction

Cherenkov light is often considered as a parasite signal in light-based dosimetry such as with scintillators [1] but it has been recently demonstrated that it can be used as the primary signal for dosimetry using polarized imaging [2, 3]. Cherenkov dosimetry has several benefits, one of which making real-time radiation treatment dosimetry possible [4]. As such, this dosimetry method is worth investigating. Cherenkov light is polarized by nature, however, the Cherenkov signal produced in a water phantom and captured by a camera is not entirely polarized due to scattering. To correct this, Cloutier et al. developed a method based on the Malus law ($I = I_{pol}(\cos(\alpha_0 - \alpha))^2 + I_{rand.pol}$) [5] and Monte Carlo simulations to extract the polarized portion of the light (I_{pol}) captured by a camera and calculate angular correction distributions to have a signal proportional to dose [2]. Such distributions account for Cherenkov photon loss due to photon interactions with the environment. As the distance between the Cherenkov source (i.e. the water phantom) and the camera increases, the cone of Cherenkov light captured by the camera becomes narrower. As a result, there is interdependencies between the geometrical setup and the Cherenkov polarization [6] reaching the camera. A connection between I_{pol}, angular correction distributions and the Cherenkov source to camera distance exists. The objective of this work is to experimentally characterize and quantify the effects of the Cherenkov source to camera distance on I_{pol} as to avoid calculating the angular correction distributions from Monte Carlo simulations.

Journal of Physics: Conference Series **2799** (2024) 012003 doi:10.1088/1742-6596/2799/1/012003

2. Materials and methods

Extractions of the purely polarized signals captured by a camera (I_{pol}) is based on Cloutier et al.'s method [2, 3]. To do so, a 15×15×20 cm^3 water tank was irradiated by a 5×5 cm^2, 6 MV photon beam (Truebeam, Varian Medical Systems, Palo Alto, USA) at a 600 MU/min dose rate for 2 minutes. A CCD camera (Atik 414EX; Atik Cameras, Norwich, United Kingdom) was placed at 1 m, 2 m, 3 m and 4 m from the water tank with its optical axis aligned with the center of the water tank to capture images of its Cherenkov emissions. Image acquisitions lasted 15 s, resulting in approximatively 6 usable images per irradiation. A polarizer was placed in front of the camera. Acquisitions were taken at 4 angles of polarization (0°, 45°, 90° and 135°) per camera placement. After subtracting the background and averaging every image of a given irradiation, a regression following Malus's law [5] was then applied to each pixel to extract I_{pol}. An alternate method based on scintillators instead of Cherenkov emissions was also used. The scintillation light captured by the camera being directly proportional to dose [2], the scintillator to camera distance is expected to not have as much impact on the projected dose distributions as the Cherenkov signal. This is due to the scintillation signal not being naturally polarised and so, not needing angular correction distributions, as opposed to the Cherenkov signal. As such, the scintillation signal was used as reference for the dose distributions. Images of the scintillation light emitted by a 10×10×10 cm^3 polyvinyltoluene-based plastic scintillator volume (EJ-260: n = 1.58, peak emission of 492 nm, density of 1:023 g/cm^3; Eljen technology, Sweetwater, TX) irradiated identically as the water tank were taken. Figure 1 shows an example of one of the set ups used. Projected percent depth doses (PPDDs) calculated with the Cherenkov I_{pol} signal from each water tank to camera distance were compared against each other and to results obtained using the scintillation signal. These comparisons are based on their relative differences and standard deviations to one another. At depths deeper than 10 cm, the Cherenkov-based results were only compared against each other since the scintillation volume has a maximum depth of 10 cm (versus 20 cm for the water tank). Uncertainties for the signal source-to-camera distance are based on the camera and its stand's dimensions. The standard deviation on pixel intensities for bands of a certain width are used as PPDDs uncertainties. These are propagated when calculating relative differences.

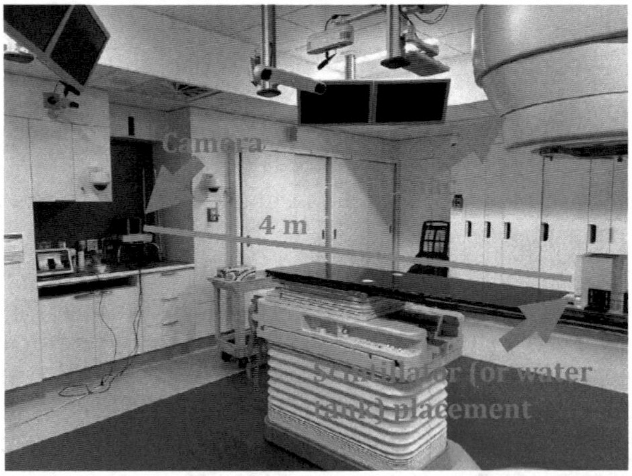

Figure 1. Picture of the set up used for the imaging of the irradiation of a scintillator (or a water tank) with a signal source to camera distance of 4m .

3. Results

Projected percent depth dose from the scintillation signal acquired at 2 m, 3 m and 4 m from the scintillator cube have an average standard deviation to each other of 0.6%. The average standard deviation between each PPDD were calculated by first calculating the standard deviation between each curve at the same depths, then averaging these standard deviations. Such standard deviation justifies the use of an average projection percent depth dose calculated from the scintillation signals as the overall scintillation PPDD for later comparisons. The PPDDs from the scintillation and from the Cherenkov I_{pol} signals are shown in figure 2. Relative differences between Cherenkov I_{pol} PPDDs and the average scintillation PPDD were extracted and are shown in figure 3. The average relative differences for each curve in figure 3 are respectively -11%, -4.8%, -2.8% and -1.6% for the 1 m, 2 m, 3 m and 4 m water tank to camera distances. The Cherenkov I_{pol} PPDD at a 4 m water tank to camera distance is the closest to the average scintillation PPDD with an average relative difference of -1.6% and an average standard deviation on this relative difference of 1.6%. Figure 4 presents relative differences between the Cherenkov-based PPDDs versus the water tank to camera distances at fixed depths in the PPDD.

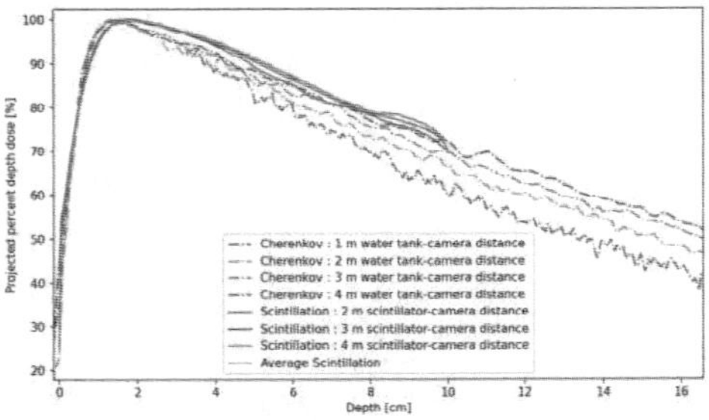

Figure 2. Projected percent depth dose calculated from the images acquired using the Cherenkov I_{pol} signal and the scintillation signal at different light source to camera distances.

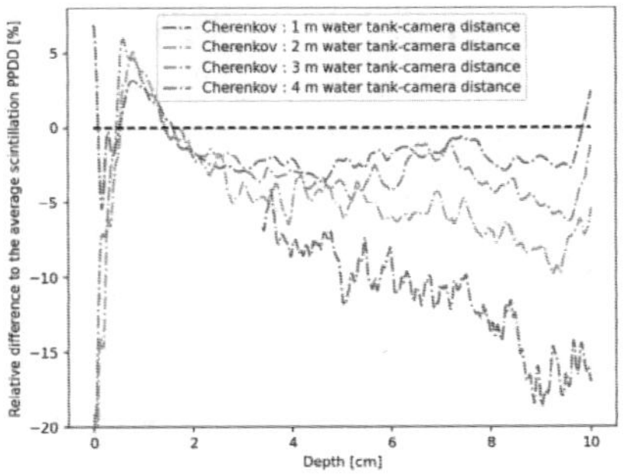

Figure 3. Relative differences along the depth (parallel to the beam axis) between Cherenkov I_{pol} PPDDs and the average scintillation PPDD.

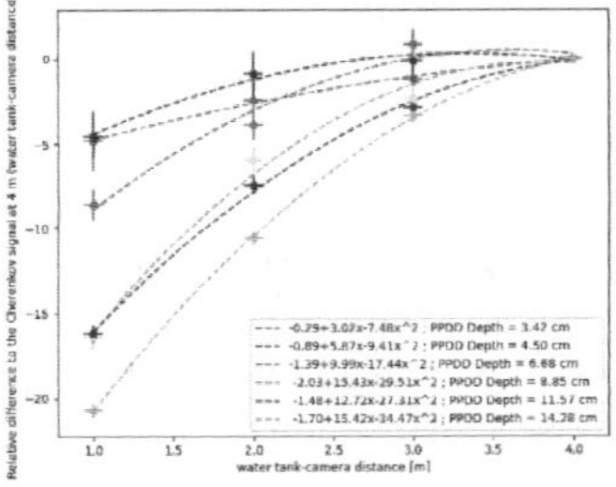

Figure 4. Relative differences between the Cherenkov I_{pol} PPDD acquired for the water tank to camera distance of 4 m and the other Cherenkov I_{pol} PPDDs at fixed depths.

Journal of Physics: Conference Series **2799** (2024) 012003 doi:10.1088/1742-6596/2799/1/012003

4. Discussion

A decrease in relative difference between PPDDs (average scintillation versus Cherenkov-base) with the increase of water tank to camera distance was observed. Indeed, there is a -11% relative difference when the camera is placed 1 m from the water tank versus a -1.6% relative difference when placed 4 m from it. Quadratic functions are fitted to the relative difference points in figure 4 to quantify the observation that as the camera is placed further from the Cherenkov light source, dose distributions require fewer angular correction distributions. This is caused by the diminution of the solid angle captured by the camera when it is placed further from the Cherenkov light source. The diminution of this solid angle results in the optical system (camera and polarizer) capturing a quantity of polarized light that is less dispersed along the optical axis. The reasoning behind the choice for the use of quadratic functions in Figure 4 is that the solid angle captured by the camera decreases with distance following the inverse square law, creating a dependency to an r^2 factor. Quadratic functions allow to consider the r^2 factor to the first order and other factors that come into play. Furthermore, the need for corrections increases with the PPDD depth as can be seen with the steepening of the fitted curves at deeper positions in figure 3. This is due to the beam being attenuated by the water phantom, which results in less Cherenkov production deeper into the PPDD.

5. Conclusion

In this work, polarized imaging of Cherenkov emissions from a water tank as well as scintillation signals from a cubic scintillator were used to calculate projected depth dose distributions. The distance from the camera to the signal source was varied and the effects on the PPDDs produced by these changes were characterized. It can be deduced from the average relative difference decreasing with water tank to camera distance (-11% at 1 m and -1.6% at 4 m) that placing the camera further from the Cherenkov light source results in the PPDD requiring less angular corrections. It shouldn't be ignored that there are practical limitations to placing the camera further from the signal source to reduce the angular correction distributions in routine radiotherapy practice. One of them being the size of the room not always being large enough to accommodate distances over 3.5 m between the camera and the signal source. As a result, choosing an appropriate distance balancing image quality and signal source to camera distance would lessen the need for Monte Carlo simulations to extract angular correction distributions.

References

[1] Beaulieu L and Beddar S, 2016. Review of plastic and liquid scintillation dosimetry for photon, electron, and proton therapy, *Physics in Medicine & Biology* **61** 20.

[2] Cloutier É, Archambault L and Beaulieu L 2022. Accurate dose measurements using Cherenkov emission polarization imaging, *Medical Physics* **49** 8.

[3] Cloutier É, Beaulieu L and Archambault L 2022. Direct in-water radiation dose measurements using Cherenkov emission corrected signals from polarization imaging for a clinical radiotherapy application, *Scientific Reports* **12** 1.

[4] Lesley A. Jarvis, Rongxiao Zhang, David J. Gladstone, Shudong Jiang, Whitney Hitchcock, Oscar D. Friedman, Adam K. Glaser, Michael Jermyn, and Brian W. Pogue, 2014. Cherenkov Video Imaging Allows for the First Visualization of Radiation Therapy in Real Time. *Int J Radiat Oncol Biol Phys* **89** 3.

[5] Collett E. Field Guide to Polarization. SPIE; 2005.

[6] Doering M, Bernloehr K, Hermann G, Hofmann W, Lampeitl H. Measurement of the Cherenkov light spectrum and of the polarization with the HEGRA-IACT-system. In: International Conference on Rebooting Computing Proceedings. 2001.

Journal of Physics: Conference Series **2799** (2024) 012004 doi:10.1088/1742-6596/2799/1/012004

On the validity of using spectrophotometry of cuvettes to provide a dose-response calibration for radiochromic optical computed tomography-based three-dimensional dosimetry

M B Jensen[1,2,3], P Balling[4,5], S H Vindbæk[1], S J Doran[6], M L Jensen[4], J B B Petersen[3], L P Muren[1,2]

[1] Danish Centre for Particle Therapy, Aarhus University Hospital, Aarhus, Denmark
[2] Department of Clinical Medicine, Aarhus University, Aarhus, Denmark
[3] Department of Medical Physics, Aarhus University Hospital, Aarhus, Denmark
[4] Department of Physics and Astronomy, Aarhus University, Aarhus, Denmark
[5] Interdisciplinary Nanoscience Center, Aarhus University, Aarhus, Denmark
[6] The Institute of Cancer Research, London, UK

E-mail: morten.bjoern.jensen@rm.dk

Abstract. The dose response of materials used in radiochromic three-dimensional (3D) dosimeters is often characterised via spectrophotometry of small cuvette samples as it is less time-consuming and cheaper. However, spatially-resolved dose measurements for clinical purposes are obtained using 3D optical computed tomography (CT). Hence, the link between the method of characterisation and the method for extracting 3D dose information must be established. The aim of this study was to investigate if spectrophotometry-based dose-response characterisation using cuvette-sized samples is valid for calibration of 3D optical CT readout systems using a silicone-based radiochromic dosimeter. Additionally, the uniformity of 3D optical CT dose-response readout of larger samples was investigated. We found the dose response of cuvette-sized samples read out by spectrophotometry to agree with the averaged dose response across cuvette-sized samples read out by optical CT and with the response near the edges of larger samples scanned using the same apparatus. Thus, cuvette-based dose-response spectrophotometry can be used to calibrate 3D optical CT readout using silicone-based radiochromic dosimeters. However, caution should be taken when considering larger samples as they exhibit a gradual radial increase in dose response from the centre and outwards.

1. Introduction

The increased complexity of radiotherapy treatment-delivery techniques has emphasised the need for a three-dimensional (3D) dosimetry system with high spatial resolution. Ideally, dosimeter characterisation should be performed using the same readout modality used to extract the 3D dose information for clinical purposes. Nonetheless, the dose response of materials used in 3D radiochromic dosimetry is often characterised via point-measurement spectrophotometry of small cuvette samples [1,2] as it is less time-consuming and cheaper. By contrast, the aim of the research field is to make spatially-resolved dosimetry measurements using 3D optical computed tomography (CT). Hence, the transition from cuvette-sized samples read out by spectrophotometry to larger samples read out by 3D optical CT must be investigated, and dose responses between the two approaches should be compared [3]. The aim of this study was therefore to investigate if spectrophotometry-based dose-response characterisation of cuvette-sized samples can be used to calibrate 3D optical CT readout systems using silicone-based radiochromic dosimeters [4,5]. Additionally, the uniformity of the 3D optical CT dose response for larger samples was assessed.

2. Materials and methods

2.1. Fabrication of dosimeters
Two batches of dosimeters were fabricated following a previously established protocol [2] from 93.24% (w/w) silicone elastomer (SE), 5.00% (w/w) curing agent (CA), 1.50% (w/w) chloroform (Sigma-Aldrich) and 0.26% (w/w) leucomalachite green (LMG - Sigma-Aldrich). The SE and CA came from the Sylgard 184 silicone elastomer kit (Dow). The LMG was dissolved in chloroform and thoroughly mixed with SE and CA using an electric mixer before air bubbles were removed using a vacuum desiccator. The mixture was moulded and left to cure, protected from light, in a chemistry fume hood, which circulated air from the outside, exposing the dosimeters to temperatures of the season (December). Moulds used were polystyrene (PS) cuvettes (VWR - 1 x 1 x 4.5 cm^3) and two types of in-house made aluminium cylinders (inner diameter = 5.0 cm; height = 5.0 cm - inner diameter = 7.5 cm; height = 7.5 cm). Aluminium moulds were sprayed with Releasil A silicone spray (Dow) before the mixture was poured into it (to ease demoulding). Batches were fabricated 72 h before irradiation (all times stated are approximate). For Batch 1, dosimeters were demoulded 5 h before irradiation, whilst for Batch 2, they were demoulded 19 h before irradiation (to save time on the day of the experiments). In both batches, some cuvette-sized samples were left in the PS container, referred to as *contained*, whereas those removed from the container (to eliminate reflection and refraction of light at the PS walls for optical CT readout) will be referred to as *uncontained* cuvettes.

2.2. Pre- and post-irradiation optical readout
Samples were read out optically before (pre-scan) and after (post-scan) irradiation (4 h per session). *Contained* cuvettes were read out using the Spectroquant Pharo 100 spectrophotometer (Merck KGaA) at 633 nm (tech specs: centre wavelength 633 nm ± 1 nm; spectral bandwidth 4 nm). *Uncontained* cuvettes and cylindrical dosimeters were read out using the Vista 16 optical CT scanner (Modus Medical Devices) with a wavelength of 632.8 nm (tech specs: centre wavelength 632.8 ± 2 nm; full width half maximum 10 ± 2 nm) following established code of practice [3,6]. The spectra were verified in-house using the LR2 (Lasertack GmbH) spectrometer. Samples were loaded into the rotary stage of the optical CT scanner using in-house 3D-printed grabbers made of polylactic acid (PLA). A complete substitution of refractive index (RI) matching liquid in the scanner was carried out between the two batches.

2.3. Irradiation setup and conditions
The dosimeters were centred in a filled acrylic glas water tank (30 x 42 x 19 cm^3) on a 3D-printed PLA platform. The setup was CT scanned (Brilliance CT BigBore, Philips Medical Systems) and imported to the Eclipse Treatment Planning System (TPS - version 16.1 - Varian Medical Systems). Treatments were planned with two opposing 18 cm x 18 cm jaw-defined flattened photon fields and normalised to deliver the prescribed dose at the isocenter, which was chosen to coincide with the centre of mass of the dosimeters. The treatment plans were designed to deliver as uniform a dose distribution as possible throughout the samples. The setup was positioned in the treatment room using in-room lasers and irradiated using a Varian TrueBeam accelerator (Varian Medical Systems) at Aarhus University Hospital, Aarhus, Denmark. The TPS gave the number of monitor units needed to deliver a specific dose (Table 1), and the machine output factor was within the clinical tolerance for patient treatments.

2.4. Data analysis
For *contained* cuvettes, the optical attenuation coefficient (μ) was calculated as the difference in absorbance as measured by spectrophotometry between post- and pre-scan multiplied by ln(10)/[cm]. For dosimeters read out by optical CT, 3D maps of μ were reconstructed with 500 projections using the ordered subset convex with total variation algorithm provided by the Vista 16 software with a resolution of 1 x 1 x 1 mm^3. Attenuation coefficients were averaged over 375 (5 x 5 x 15) voxels in the bulk of *uncontained* cuvettes and over 1000 (10 x 10 x 10) voxels at the centre of the cylinders. Near the edge of the cylinders, μ was averaged over a circular ring with an inner radius of 1.8 cm and outer radius of

2.2 cm for the *ø 5.0 cm* samples (1092 voxels) and an inner radius of 3.0 cm and outer radius of 3.4 cm for the *ø 7.5 cm* samples (1596 voxels). The dose responses, including 95% confidence intervals (CI), were given as the slope obtained by fitting a linear regression to μ (*ø 7.5 cm centre* receiving 12 Gy was omitted due to pronounced *cupping* [7]) plotted against dose using STATA (StataCorp). For Batch 2, zero-dose reference *contained* and *uncontained* cuvettes were brought back and forth to the irradiation facility under the same conditions as the other dosimeters and read out simultaneously, and the optical responses of *contained* and *uncontained* cuvettes were adjusted by subtracting the respective reference values. Radial cross-sectional line profiles were obtained by averaging 49 voxels (7 x 7) through the middle of the cylinders and normalised to the centre value.

Table 1: Samples, readout method, dose levels and number of samples per dose level for Batch 1 and 2.

samples	readout method	Batch 1		Batch 2	
		dose levels (Gy)	Samples per dose level	dose levels (Gy)	Samples per dose level
contained cuvettes	spectrophotometer	3,6,9,12	3	0,2,4,6,8	2
uncontained cuvettes	optical CT	3,6,9,12	3	0,2,4,6,8	2
ø 5.0 cm	optical CT	3,6,9,12	1	2,4,6,8	1
ø 7.5 cm	optical CT	3,6,9,12	1	2,4,6,8	1

3. Results

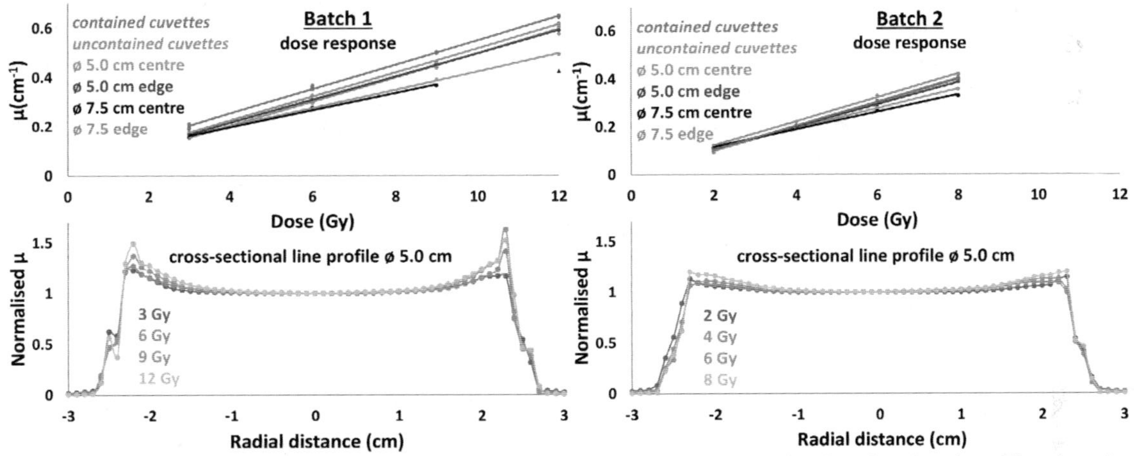

Figure 1: Left column: Batch 1. Right column: Batch 2. Top row: linear regressions of μ plotted against dose. The triangular data point was excluded from the linear regression. Bottom row: normalised cross-sectional line profiles.

All dose responses showed linearity (top panels of Figure 1). Disregarding *ø 7.5 centre* from Batch 1 due to low statistics (only three data points), the responses in the central regions of the cylindrical dosimeters were found to agree (overlapping CIs) and to be lower compared to the rest of the responses, which agreed with each other (Table 2). Non-uniform cross-sectional line profiles were observed for the cylinders (bottom panels of Figure 1). For Batch 2, the average response of zero-dose reference of *contained* cuvettes was 0.0253 $cm^{-1,}$ while for *uncontained*, it was 0.0023 cm^{-1}.

Table 2: Slope (dose response) and intercept for linear regressions including 95% CIs. $R^2 > 0.99$ for all regression models.

samples	readout method	Batch 1		Batch 2	
		dose response	intercept	dose response	intercept
contained cuvettes	spectrophotometer	$(49.1 \pm 1.4)\cdot10^{-3}$ $cm^{-1}Gy^{-1}$	$(57.6 \pm 11.8)\cdot10^{-3}$ cm^{-1}	$(49.4 \pm 0.8)\cdot10^{-3}$ $cm^{-1}Gy^{-1}$	$(6.9 \pm 4.6)\cdot10^{-3}$ cm^{-1}
uncontained cuvettes	optical CT	$(49.0 \pm 1.9)\cdot10^{-3}$ $cm^{-1}Gy^{-1}$	$(7.3 \pm 15.6)\cdot10^{-3}$ cm^{-1}	$(49.7 \pm 2.2)\cdot10^{-3}$ $cm^{-1}Gy^{-1}$	$(0.6 \pm 12.1)\cdot10^{-3}$ cm^{-1}
ø 5.0 cm centre	optical CT	$(36.8 \pm 4.1)\cdot10^{-3}$ $cm^{-1}Gy^{-1}$	$(54.5 \pm 33.6)\cdot10^{-3}$ cm^{-1}	$(40.9 \pm 2.4)\cdot10^{-3}$ $cm^{-1}Gy^{-1}$	$(31.0 \pm 13.1)\cdot10^{-3}$ cm^{-1}
ø 5.0 cm edge	optical CT	$(46.8 \pm 1.3)\cdot10^{-3}$ $cm^{-1}Gy^{-1}$	$(28.7 \pm 10.8)\cdot10^{-3}$ cm^{-1}	$(46.2 \pm 4.6)\cdot10^{-3}$ $cm^{-1}Gy^{-1}$	$(16.1 \pm 25.1)\cdot10^{-3}$ cm^{-1}
ø 7.5 cm centre	optical CT	$(34.6 \pm 31.9)\cdot10^{-3}$ $cm^{-1}Gy^{-1}$	$(59.1 \pm 206.8)\cdot10^{-3}$ cm^{-1}	$(36.2 \pm 6.9)\cdot10^{-3}$ $cm^{-1}Gy^{-1}$	$(44.8 \pm 37.6)\cdot10^{-3}$ cm^{-1}
ø 7.5 cm edge	optical CT	$(48.6 \pm 3.6)\cdot10^{-3}$ $cm^{-1}Gy^{-1}$	$(31.0 \pm 29.6)\cdot10^{-3}$ cm^{-1}	$(49.7 \pm 4.8)\cdot10^{-3}$ $cm^{-1}Gy^{-1}$	$(24.2 \pm 26.4)\cdot10^{-3}$ cm^{-1}

4. Discussion

The dose responses of *contained* cuvettes read out by spectrophotometry concurred well with *uncontained* cuvettes read out by optical CT.

The agreement of dose responses across the two batches indicated good reproducibility of dosimeter fabrication. Dose responses of *contained* cuvettes found in this study were higher than those found in a previous study with the same chemical composition and curing time [2] and are likely attributed to different curing temperatures, as experiments in the previous study were conducted in May.

When the total absorption through radiochromic dosimeters becomes too high, the optical CT algorithms fail to reconstruct the actual μ distribution, typically predicting an artificially low absorption in the central regions, denoted as *cupping* [7]. In Batch 2, we tried to reduce this effect by reducing the doses and adding more dye to the RI matching liquid (i.e., increasing dynamic range and exposure time) [3]. However, further investigations are needed to determine the impact of *cupping* on larger samples.

A combination of four effects causes the non-uniform dose responses observed in the cross-sectional line profiles: (i) imperfect sample positioning between pre- and post-scan, (ii) imperfect RI matching liquid, (iii) non-uniform sensitivity to dose within the sample and (iv) *cupping* [8]. The first two account for the abrupt changes and peaks at the very edge of the dosimeters, whilst the last two account for the gradual increase of response to dose from the centre and outwards. We tried to eliminate the first two effects for Batch 2 by complete RI matching liquid substitution and optimising the method of sample loading into the scanner; the improvement is evident (bottom panels of Figure 1). To our knowledge, this is the first time that edge effects have been reported for the silicone-based radiochromic dosimeter. However, similar effects have previously been reported for another radiochromic dosimeter and methods of correction have been proposed [8].

This study indicates that moulds and removal of such before irradiation affect the dose sensitivity and creep-up effect (non-irradiation-induced increase in μ) [1]. The effect is observed in the zero-dose references, where the difference in response between the *contained* and *uncontained* samples is believed to be caused by the PS container rather than by distinct readout modalities. The difference in responses in the centre of cylinders between the two batches could be caused by different demoulding times. Potentially, early demoulding could reduce edge effects, however, further investigations are needed.

Limitations of this study include different demoulding times and dose levels across batches and a low sample size. The sample size was balanced against an increased time interval between irradiation and readout as this time interval affects creep-up and fading (decrease in μ after irradiation) [1]. Scanning and irradiation took approximately 12 h, and samples were stored in a cooler to increase temporal stability [9].

In conclusion, we found that the dose response of cuvette-sized samples read out by spectrophotometry agreed with the averaged dose response across cuvette-sized samples read out by optical CT and with the response near the edges of larger samples scanned with the same apparatus using silicone-based radiochromic dosimeters. Thus, cuvette-based dose-response spectrophotometry can calibrate 3D optical CT readout using silicone-based radiochromic dosimeters. However, caution should be taken when considering larger dosimeters as they exhibit a gradual radial increase in dose response from the centre and outwards.

5. Acknowledgements
This research was funded by the Novo Nordisk Foundation (Grant number NNF18OC0034718).

6. References
[1] Du Y *et al* 2020 *Radiat. Phys. Chem.* **168** 108574
[2] Jensen M B *et al* 2022 *Acta Oncol.* **61** 264–8
[3] Doran S J 2010 *J. Phys. Conf. Ser.* **250** 012086
[4] De Deene Y *et al* 2015 *J. Phys. Conf. Ser.* **573** 012025
[5] Høye E M *et al* 2015 *J. Phys. Conf. Ser.* **573** 012067
[6] Doran S J 2013 *J. Phys. Conf. Ser.* **444** 012004
[7] Al-Nowais S *et al* 2009 *J. Phys. Conf. Ser.* **164** 012023
[8] Costa F *et al* 2020 *Phys. Med. Biol.* **65** 095003
[9] Valdetaro L B *et al* 2023 *Med. Phys.* **50** 2560–4

Journal of Physics: Conference Series **2799** (2024) 012005 doi:10.1088/1742-6596/2799/1/012005

Mapping 3D doses in water with a cable robot equipped with plastic scintillator

Louis Archambault[1,2], Ramin Mersi[3], Simon Foucault[3] Frédérik Berthiaume[4], Boby Lessard[4], Francois Therriault-Proulx[4] and Philippe Cardou[3]

[1]Department of physics, Université Laval, Québec, Canada

[2]Oncology division, CHU de Québec-Université Laval Research Center, Québec, Canada

[3]Department of mechanical engineering, Université Laval, Québec, Canada

[4]Medscint inc., Québec, Canada

E-mail: louis.archambault@phy.ulaval.ca

Abstract. This work presents the first use of a cable robot for 3D dosimetry. Its design was optimized to operate in water and offer five degrees of freedom. It was equipped with a plastic scintillation detector. The cables and the end effector were made of plastic, thus making all components inside the phantom water equivalent. Feasibility and reproducibility was demonstrated using a 6MV beam. The cable robot prototype had a usable workspace of 40 cm in each direction and could rotate over 90 degrees in both the polar and azimuthal directions. It was tested at speeds of up to 120.8 mm/s. Position precision was within 2.5 mm for translation and within 1.1 mm for rotations about a fixed point. This new robotic dosimetry system is more versatile and can address limitations of current water phantoms that are constrained to only 3 translation axes.

1 Introduction

In-water dose measurements are essential for accurate dose calculation in radiation therapy as those measurements are used for commissioning and quality assurance of linear accelerators and treatment planning systems. Motorized, computer controlled water tanks for dosimetry have been around for over 50 years [1]. A dosimeter attached to such system can be scanned to acquire dose profiles along a given axis of motion. Motorized water phantoms are usually rectangular to allow scans along any of the three cartesian axes. Their design remained relatively similar over time with the exception of the system developped by Sun Nuclear (Sun Nuclear Corporation, FL, USA) that adopted a cylindrical shape. The rationale for this was to be capable of acquiring profiles at any angle in the plane perpendicular to the beam axis without having to rotate the tank [2]. Such profiles are important for treatment modalities that uses circular fields such as iris-based Stereotactic Radiosurgery (SRS). Comparison of different commercial motorized water phantom have shown that they can all achieve equivalent level of accuracy: dose difference within 1% and distance to agreement within 1 mm of each other [3]. Nevertheless, these phantoms do suffer from drawbacks; their large metallic arms can interfere with the radiation beam [4] and can cause waves that lead to measurement innacuracies [5] of up to 5% [6]. Risks of interference with the metallic arm usually restrict irradiations at normal incidence (i.e. 0 degree gantry angly). In that context, we developed a new robotic dosimetry system that uses a 5 degree of freedom (5DOF) robot to move a plastic scintillation detector (PSD) within a water phantom. Motion along 3 spatial (x, y, z) and 2 angular (ϕ: azimuthal; θ: polar) directions are possible and can be combined to move the detector along trajectories that are not restricted to be parallel to one the three cartesian directions. Furthermore, because of the angular DOFs, it is possible to alternate rapidly between scans where the detector is parallel to the beam axis to scans where the detector is perpendicular to the beam axis. Because the cables are small and polymer based, they do not cause beam interference. Thus, this new system is ideal for mapping complex 3D doses.

Journal of Physics: Conference Series **2799** (2024) 012005 doi:10.1088/1742-6596/2799/1/012005

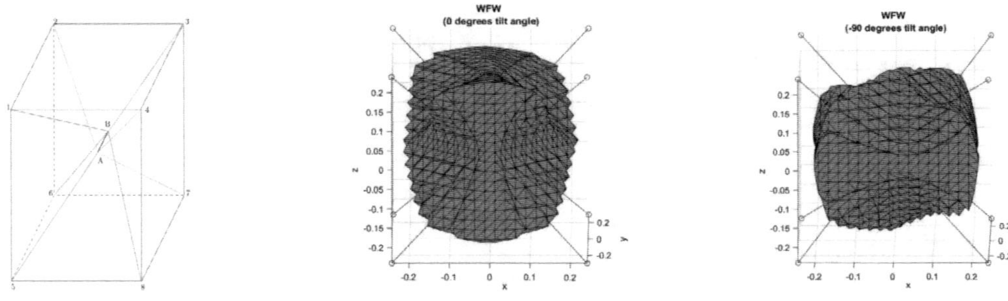

(a) (b) (c)

Figure 1: (a) The selected configuration for the 8 cable position. The end effector is shown in black in the center of the water tank in red. (b) the usable workspace with the orientation at $\theta = 0°$. (c) the usable workspace with the orientation at $\theta = 90°$ (all units are in meters)

2 Methods

2.1 Cable robot design

Cable robots are a type of parallel robots that have found numerous applications in different types of industry [7], but never in medical physics. This new dosimetry cable robot was designed to fit within a DoseView 3D water tank (Standard Imaging, Wisconsin, USA) which have a usable volume of $645 \times 654 \times 523$ mm^3. Because PSDs are lightweight and because the robot is to be operated in a water-filled tank, it was not possible to use gravity as a constraint to maintain cable tension. Thus, the robot was overconstrained by using more cables than strictly necessary in order to maintain cable tension at all time which is necessary for accurate movement. An eight-cable design was chosen with a rod-like end effector which is responsible for holding the PSD. Four cables were attached to each extremity of the end effectors. From the end effector, each cable went to a pulley placed in corners of the tank. From there the cable was connected to a whinch placed outside, on the side of the phantom. The risk of cable intereference was minimized by having each cable go to a different corner of the water tank. The PSD cable was passed toward the middle of the water tank to avoid interference with the robot's cables. There are 2^8 possible cable configurations. An algorithm was therefore written to find the configuration that lead to the largest workspace while guaranteeing at least 90° range of rotation in the polar and azimuthal directions (see Fig. 1). This algorithm simulated cable tensions and performed a grid-based search of possible configurations for multiple detector orientation. Cables were made of Vectran coated in polyester, a material known for its negligible deformation and resistance to environmental factors. Therefore, all components in water (cables, end effector, PSD) are made of water-equivalent polymers. The complete prototype is shown on Fig. 2. Control of the robot was achieved using RT-LAB software (Opal-RT technologies, Montreal, Canada). Positions along a given trajectory are converted from spatial cartesian coordinates to cable lengths; cable speeds are obtained by multiplying the cartesian velocities with the robot's Jacobian matrix that relates the velocities of the robot's end-effector to the velocities of its cables.

2.2 Plastic scintillation dosimeter and irradiations

The PSD attached to the end effector of the robot is a Medscint (Medscint inc. Quebec, Canada) probe (model PRB-0019) with 1 mm diameter, 1 mm long sensitive region. The overall diameter is 1.4 mm given the protective jacket. It was read with an HS-RP200 spectrometer and calibrated according to the manufacturer's specifications. All irradiations were done using a 6MV Truebeam linac (Varian Medical System, Palo Alto, USA) with a flattening filter.

3 Results

The robot could travel over 450 mm in each direction, which is slighlty less than the original Doseview 3D scanning dimensions ($480 \times 480 \times 410$ mm^2). Speeds of up to were 120.8 mm/s were achieve without producing waves that could influence dose measurements while the detector was 1 cm or more below the surface. An example of scans along crossline, inline and depth is shown along with the robot position in (x, y, z) as a function of time on Fig. 3a. This data was acquired in a single, continuous run. An example dose and position reproducibility is shown on Fig. 3b for half-profiles of a

Journal of Physics: Conference Series **2799** (2024) 012005 doi:10.1088/1742-6596/2799/1/012005

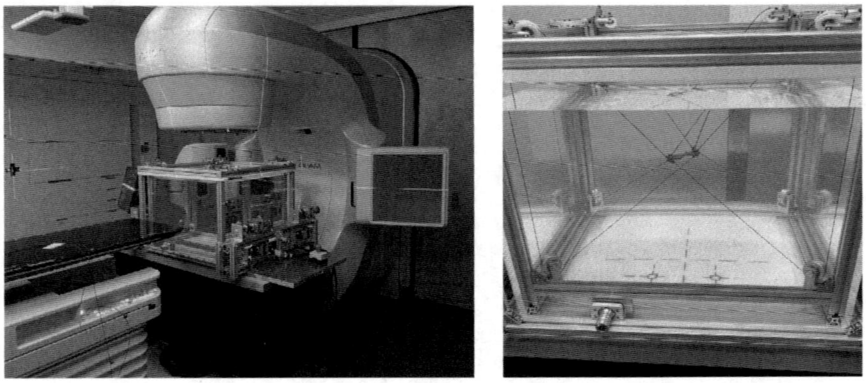

Figure 2: The cable robot

5×5 cm^2 fields. Using the location of the 50% dose level as a reference point, the precision of the robot was evaluated to be 2.5 mm. Measurements with the detector fixed while its orientation was continoussly changed along the polar (θ) and azimuthal (ϕ) angles are shown in Fig 4. Absolute dose deviation of about 2% can be explained in part by the detector angular dependence (~1.5% according to the manufacturer) and small positional drifts ($\pm$1.1 mm) when the detector is rotated. These drifts are determined from the change in cable lengths during rotations. The robot manage to achieve uniform rotation for both angular degree of freedom except a small unplanned 3° deviation in θ around 105 s.

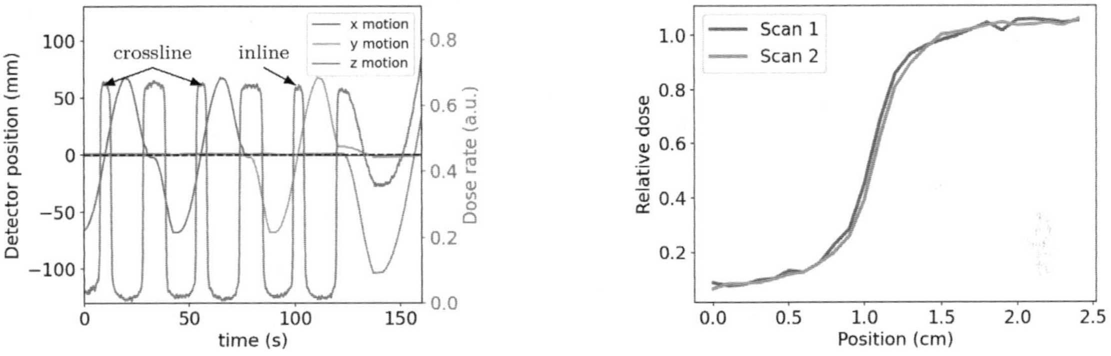

(a)

(b)

Figure 3: (a) scans of a 5×5 cm^2 field, along all three cartesian directions (x is crossline, y is inline and z is depth). The region of high doses between scans corresponds to the robot repositioning for the next scan. (b) repeated scans of the edge of a 5×5 cm^2 field to illustrate.

4 Discussion

While the position precision of this 5DOF cable robot is encouraging, it is still below that of traditional motorized water phantoms. Nevertheless, we've shown that this new device possesses multiple advantages for 3D dosimetry and possesses a usable workspace that can cover most clinical dose distributions. Because the movement can be programmed along any trajectory, it could be used to map dose distributions from non-coplanar treatments; furthermore, there is no risk of interference between the radiation beam and metallic components because the entire measurement volume is composed of water equivalent material. This absence of metal could also make this device well suited for use in magnetic resonance (MR) guided radiation therapy (MRgRT), but some modifications will be needed to make sure the motors and winches are either MR-compatible or situated at a safe distance from the regions of high MR fields. The two angular degrees of freedom is also a unique feature that can serve

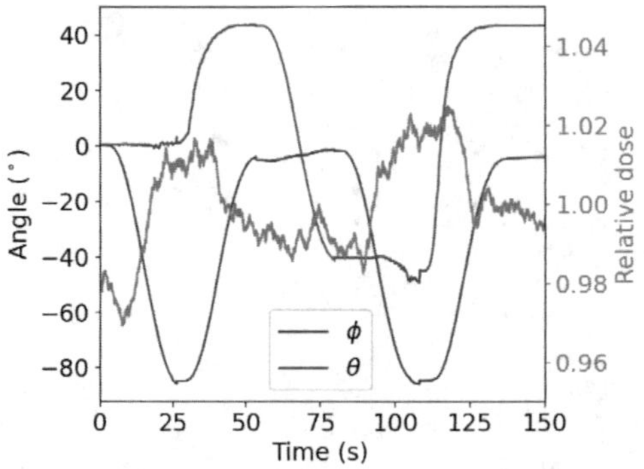

Figure 4: Angular scans along the polar (θ) and azimuthal (ϕ) angles. For these scans the detector remain fixed while the stem rotate

multiple purposes. It can be used to measure the angular dependence of a given radiation dose detector. Once this is known, the orientation of a detector can be actively controlled during measurements to minimize the impact of its angular dependence. Furthermore, for detectors such as PSDs that can suffer from stem irradiation, controlling the angle of the detector may help limit the level of stem signal. While promising, this first prototype could be improved in multiple ways. Position accuracy could be improved with better calculations that account for the presence of the pulleys that have been neglected so far. Precision and accuracy could also be improved by including an independent motion tracking system such as a computer vision system that can provide real-time feedback.

5 Conclusion

This work is the first report on the feasibility of using a cable robot for dose measurements in water. This 5 degrees of freedom robot can move a detector inside a water phantom with great versatility making it well suited for 3D dose measurements of complex or non-standard dose distribution. Future work will focus on better assessment and improvement of the accuracy of the robot.

References

[1] Smith V and Clampitt S 1974 *Physics in Medicine & Biology* **19** 244–244

[2] Al Kafi M A, Mwidu U and Moftah B 2015 *Applied Radiation and Isotopes* **105** 88–91

[3] Akino Y, Gibbons J P, Neck D W, Chu C and Das I J 2014 *Journal of appliedd clinical medical physics* **15** 251–258

[4] Aftab S, Barnes M P, Doebrich M and Lehmann J 2020 *Journal of Applied Clinical Medical Physics* **21** 95–101

[5] Das I J, Cheng C W, Watts R J, Ahnesjö A, Gibbons J, Li X A, Lowenstein J, Mitra R K, Simon W E and Zhu T C 2008 *Medical physics* **35** 4186–4215

[6] Bakhtiari M 2011 *Journal of Medical Physics* **36** 230–233

[7] Zarebidoki M, Dhupia J S and Xu W 2022 *IEEE Robotics & Automation Magazine* **29** 89–106

Journal of Physics: Conference Series **2799** (2024) 012006 doi:10.1088/1742-6596/2799/1/012006

Towards optimisation of the chemical composition of a bone-imitating dosimeter as a potential component of multiphase dosimeters

M. Jaszczak-Kuligowska[1,*], M. Kozicki[1,2,*], P. Maras[3]

[1] Department of Mechanical Engineering, Informatics and Chemistry of Polymer Materials, Faculty of Material Technologies and Textile Design, Lodz University of Technology, Żeromskiego 116, 90-543 Lodz, Poland
[2] GeVero Co., Lodz, Poland
[3] Department of Radiotherapy Planning, Copernicus Hospital, 93-513 Lodz, Poland

*E-mail: malwina.jaszczak@p.lodz.pl, marek.kozicki@p.lodz.pl

Abstract. The paper concerns a bone-imitating dosimeter called BoneGel (PAGAT2–Pluronic F–127 with hydroxyapatite). The addition of inorganic salt ($MgCl_2$) to the composition increased the dose sensitivity of the dosimeter.

1. Introduction

Over the last 30 years, research has been conducted on 3D gel dosimeters, which enable measurement of dose distribution in three dimensions with high resolution and at the same time complement modern irradiation techniques and treatment planning systems, ensuring the required accuracy and safety of radiotherapy. Several 3D dosimeter systems have been introduced over the years, such as Fricke gel dosimeters, polymer gel dosimeters, radiochromic gel dosimeters, radiofluorogenic gel dosimeters, deformable radiochromic dosimeters, and plastic gel dosimeters [1, 2]. However, only a few studies have been carried out investigating gel dosimeters mimicking other tissues than water, such as lung-imitating [3, 4, 5] and bone-imitating [6, 7] gel dosimeters. Even more rare are gel dosimeters consisting of combinations of tissue-imitating dosimeter volumes within the same phantom [8].

This work concerns a bone-imitating dosimeter (BoneGel) as a component of the multiphase dosimeter that we have developed [9], imitating lungs, soft tissues, and bones in one phantom (Figure 1). This approximation of the dosimeter to the actual anatomical structures of the human body, in which organs of different densities are adjacent to each other, may additionally increase the precision of treatment plan verification and enable the measurement of the planned dose distribution in situations where the cancer is located in soft tissues, lungs, bones, and at the border of these phases. This is of great importance due to the differences in the absorption of ionising radiation by elements of the human body with different densities, and may find application in modern radiotherapy [10]. All parts of the multiphase dosimeter are based on one PAGAT2–Pluronic F–127 gel [11], subjected to appropriate modifications (foaming to obtain a dosimeter imitating alveoli and the addition of hydroxyapatite to obtain a dosimeter imitating bone).

Towards our vision to create a multiphase dosimeter of various densities mimicking the human body, our specific aim of this study is to investigate if an addition of $MgCl_2$ can change the

Journal of Physics: Conference Series **2799** (2024) 012006 doi:10.1088/1742-6596/2799/1/012006

dose sensitivity in PAGAT2–Pluronic F–127 gel previously optimised in terms of the amount and type of hydroxyapatite [7], as observed for MAGAT [12], PAGAT, NIPAM, and VIPET [13] gels.

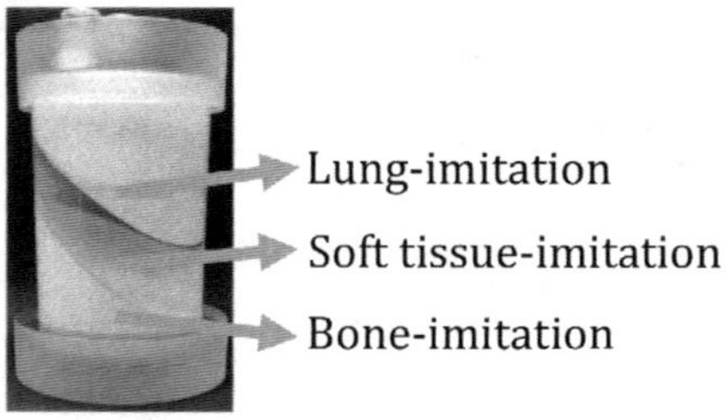

Figure 1. Photograph of a multiphase dosimeter composed of a part imitating lung alveoli, soft tissue, and bone.

2. Materials and methods

2.1 Gel preparation

The PAGAT2–Pluronic F–127 with a hydroxyapatite (BoneGel) dosimeter was composed of 25% poly(ethylene oxide)-block-poly(propylene oxide)-block-poly(ethylene oxide) (Pluronic F–127, Sigma Aldrich, USA), 3% (w/w) *N,N′*-methylenebisacrylamide (MBA, Sigma Aldrich, USA), 3% w/w acrylamide (AA, Sigma Aldrich, USA), 18% w/w hydroxyapatite (CaHA, Merz North America, USA), and 10 mM tetrakis(hydroxymethyl)phosphonium chloride (THPC, Sigma Aldrich, USA). Magnesium chloride ($MgCl_2$, Chempur, Poland) was added to the composition in two variants, 0.05 and 0.1 M. At first, MBA was dissolved in distilled water during heating to 50 °C, and it was kept at this temperature for 20 min. Subsequently, the temperature was lowered to add AA at 40 °C and $MgCl_2$ at 35 °C. Afterwards, the solution was placed in a refrigerator (4 °C) for 5 min and then mixed with a cooled down to 4 °C Pluronic F–127 33% w/w solution prepared 72 h earlier according to the procedure described elsewhere [14]. After mixing, the solution was kept at room temperature. When it reached a temperature of 14 °C, it was stirred with a mechanical mixer (about 1000 rpm) until about 17 °C and transferred to a dryer heated to 50 °C. The solution was kept in the dryer for approximately 2 hours to form a physical gel and to dissolve the MBA crystals that had not been completely dissolved. The gel was then placed in the refrigerator for approximately 15 hours, after which it returned to solution form. Then, hydroxyapatite was added to the solution in portions, stirring with a spoon and then with a mechanical mixer at low speed (about 500 rpm). Finally, THPC was added to the solution. The solution was then poured into nuclear magnetic resonance (NMR) glass vials (8.75 mm inner diameter, ~100 mm length, 0.63 mm wall thickness, sealed with rubber caps and Parafilm®) and placed in a dryer at 50 °C for 2 hours.

2.2 Irradiation

The gel samples were irradiated using a TrueBeam medical accelerator (Varian Medical Systems, USA; parameters: a photon beam of 6 MV, monitor unit rate 600 MU/min, field size 20×20 cm^2, dose range 0-40 Gy). During irradiation, the samples were submerged in a water phantom (W-1-DD1-2, GeVero Co., Poland) filled with distilled water (~21 °C) at a depth of 5 cm, SSD 95 cm.

2.3 NMR measurements

NMR measurements (20 Hz NMR relaxometer, Bruker, Germany) were performed using a Carr-Purcell-Meiboom-Gill (CPMG) pulse sequence to obtain T_2 ($R_2 = 1/T_2$). The sequence was as

Journal of Physics: Conference Series **2799** (2024) 012006 doi:10.1088/1742-6596/2799/1/012006

follows: the time delay between consecutive pulses was 10 ms; the first time-to-echo (TE) was equal to 10 ms, and 120 data points were received (TE range: 10–1200 ms). T_2 was calculated based on the first order exponential free induction decay. The measurements were performed at 25 °C (set with the aid of a heating-cooling system (Haake, Germany) and a proton-free heating-cooling liquid (Fluorinert™ FC-43, 3M Company)). The gel dosimeter samples were measured several times from 24 h to 11 days after irradiation (one vial of the dosimeter for each dose).

2.4. CT measurements

BoneGel samples with 0 and 0.5 M $MgCl_2$ in syringes (50 ml, Webest, Medica, China) were measured with a Somatom Sensation Open CT scanner (Siemens AG Medical Solutions, Germany). They were inserted into an electron density phantom (model 062, CIRS Norfolk, VA, USA) together with the insert imitating trabecular bone 200 mg/cc. The scanning parameters were as follows: 120 kV, eff. mAs 115, pitch 0.5, slice thickness 1.5 mm, field-of-view (FOV) 500 mm. The B31f reconstruction algorithm was used (the spacing between images was 1.5 mm).

3. Results and discussion

The NMR readout measurements resulted in the dose-responses of dosimeters without and with $MgCl_2$ were plotted and described using the equations in Figure 2.

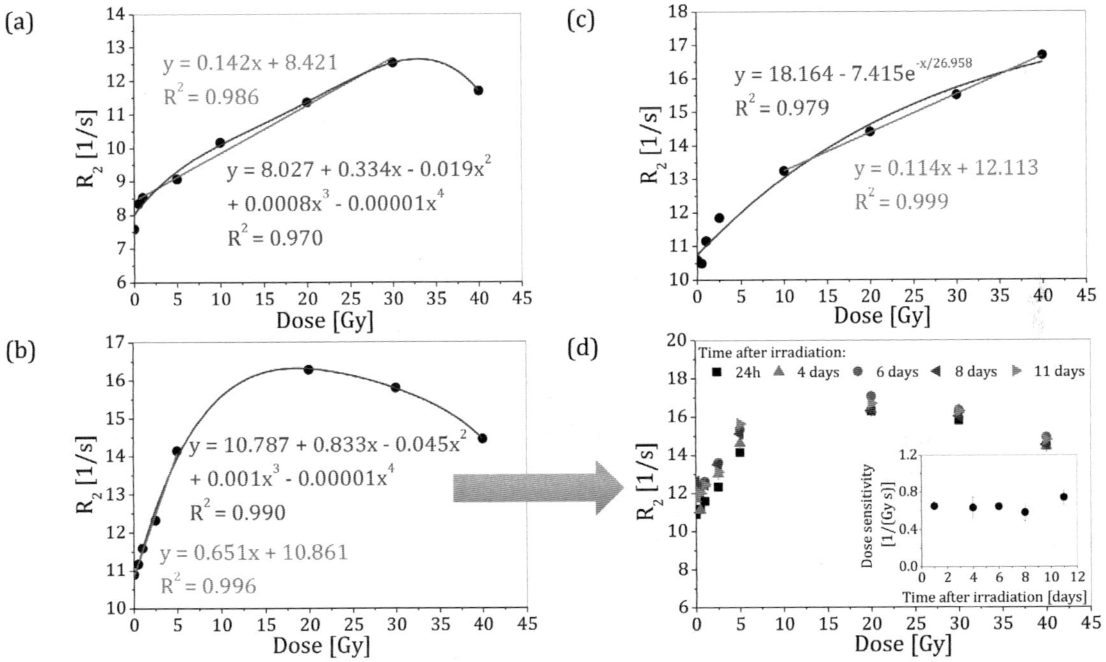

Figure 2. Dependence of the spin-spin relaxation rate (R_2) as a function of absorbed dose for BoneGel dosimeter without (a) and with the addition of 0.05 M (b) and 0.1 M (c) $MgCl_2$ with fitted curves for dynamic (blue curves) and linear (red curves) dose ranges. The stability over time for optimal composition with 0.05 M $MgCl_2$ (d). The measurements were performed 24 h after irradiation (a–c) and for 11 days after irradiation (d).

The obtained relationships allowed the determination of basic parameters of dosimeters such as dynamic and linear range, threshold dose, dose sensitivity and intercept, which are presented in Table 1. The addition of magnesium chloride to the BoneGel composition increased the sensitivity of the dosimeter similarly to other polymer gel dosimeters [12, 13], but only for the lower tested $MgCl_2$ concentration, 0.05 M (Figure 2b). In this case, a 4.6-fold increase in dose sensitivity was achieved. The larger amount of magnesium chloride in the composition (0.1 M) was difficult to dissolve and resulted in a 20% decrease in dose sensitivity (Figure 2c) and a shift in the linear range towards higher doses.

For the optimal composition (0.05 M $MgCl_2$), stability tests over time were performed by analysing the R_2 values and dose sensitivity values of the dosimeter (Figure 2d). From the results obtained, it can be concluded that the dosimeter is stable for at least 11 days after irradiation. Due to the high value of dose sensitivity and stability over time after irradiation, the composition containing 0.05 M of magnesium chloride was selected as a part imitating bone in a multiphase dosimeter in further work.

The HU values are similar for the dosimeter without $MgCl_2$ (260 HU) and with the addition of 0.5 M $MgCl_2$ (291 HU), as well as close to those for trabecular bone (239 HU).

Table 1. The basic parameters of the BoneGel dosimeters, determined on the basis of the dose-responses (Figure 2).

Concentration of $MgCl_2$ [mol]	Threshold dose [Gy]	Linear dose range [Gy]	Dynamic dose range [Gy]	Linear equation y = ax + b, where y is R_2, x is Dose, and b is $R_{2,0}$		
				Dose sensitivity (a) [Gy^{-1} s^{-1}]	Intercept ($R_{2,0}$) [s^{-1}]	R^2
0	<0.5	0.5–30	<0.5–30	0.142±0.008	8.421±0.112	0.986
0.05	<0.5	0–5	<0.5–20	0.651±0.022	10.861±0.074	0.996
0.1	1	10–40	1–40	0.114±0.001	12.113±0.035	0.999

4. Conclusions

The addition of inorganic salt ($MgCl_2$) in an appropriate amount increases the dose sensitivity of the PAGAT2–Pluronic F-127 with hydroxyapatite (BoneGel) dosimeter. In further work, this gel will be used as a bone-imitating component in a multiphase dosimeter to measure dose distribution and verify the treatment plan in radiotherapy using computed tomography as an imaging method.

Acknowledgement: The authors kindly thank Merz North America, USA, for providing the hydroxyapatite component for this study.

References

[1] Baldock C et al., 2010, Phys. Med. Biol., 55, R1–R63.
[2] De Deene Y, 2022, Gels, 8, 599.
[3] Kozicki M et al., 2021, Radiat. Phys. Chem., 185, 109479.
[4] De Deene Y et al., 2006, Med. Phys., 33, 2586–2597.
[5] Haraldsson P et al., 2006, Phys. Med. Biol., 51, 919–928.
[6] Kumahara N et al., 2023, Radiol. Phys. Technol., 16, 227–234.
[7] Kozicki M et al., 2024, Meas.: J. Int. Meas. Confed. Submitted.
[8] Kozicki M et al., 2023, Adv. Mater. Technol., 8, 2201023.
[9] Kozicki M et al., 2024, IC3DDose 2024, IOP J. Phys.: Conf. Ser. Submitted.
[10] Jensen M L et al., 2023, Medical Physics, 51, 2200–2209.
[11] Jaszczak M et al., 2018, Phys. Med. Biol., 63, 175010.
[12] Hayashi S-i et al., 2012, Radiat. Phys. Chem., 81, 884-888.
[13] Hayashi S-i et al., 2018, Radiol. Phys. Technol., 11, 375-381.
[14] Kozicki M et al., 2017, Phys. Med. Biol., 62, 5668–5690.

Journal of Physics: Conference Series **2799** (2024) 012007 doi:10.1088/1742-6596/2799/1/012007

Preliminary study on a bifunctional, elastic NBT–PVA radiochromic gel acting as a bolus and dosimeter in radiotherapy

M. Jaszczak-Kuligowska[1,*], P. Maras[2,3] and M. Kozicki[1,2,*]

[1] Department of Mechanical Engineering, Informatics and Chemistry of Polymer Materials, Faculty of Material Technologies and Textile Design, Lodz University of Technology, Żeromskiego 116, 90-543 Lodz, Poland
[2] GeVero Co., Lodz, Poland
[3] Department of Radiotherapy Planning, Copernicus Hospital, 93-513 Lodz, Poland

*E-mail: malwina.jaszczak@p.lodz.pl, marek.kozicki@p.lodz.pl

Abstract. This work presents a new elastic radiochromic gel containing a radiation-sensitive nitro blue tetrazolium chloride (NBT) embedded in a poly(vinyl alcohol) (PVA) matrix, which can simultaneously function as a bolus and a dosimeter in radiotherapy.

1. Introduction

The need to use boluses in radiotherapy is related to the characteristics of the megavolt photon beam dose distribution in the patient's body. After the beam passes through the patient's skin surface, the dose increases rapidly, reaching a maximum value, (dose build-up area), and then decreases exponentially. The presence of a dose build-up area makes it difficult to obtain a uniform dose distribution in situations where the cancer lesion is located shallowly under the skin. In such a case, a bolus, which is an additional material of appropriate thickness, can be placed on the patient's skin, which will shift the dose increment area and allow the full dose to be obtained in the therapeutic area. The bolus should be tissue-equivalent, biocompatible, and should adhere closely to the body surface. Boluses are made of various materials, such as polystyrene, acrylonitrile butadiene styrene, polycaprolactone, polylactic acid, polyurethane, silicone, and hydrogels. They are also printed for individual needs using 3D printers [1–3].

This work presents a new proposal for a bolus gel material composed of PVA and borax matrix with the addition of the radiochromic compound (NBT), thanks to which it can also function as a dosimeter for measuring the dose on patients' body surfaces. The NBT–PVA gel is flexible and transparent and changes colour from yellow to purple-brown under exposure to ionising radiation. The change is more intense the higher the radiation dose absorbed.

2. Materials and methods

2.1 Gel preparation

The NBT–PVA gel is composed of water, 3.2% (w/w) poly(vinyl alcohol) (PVA, Mw 85000–124000, Sigma-Aldrich, USA) with 0.8% (w/w) di-sodium tetraborate decahydrate (borax, Chempur, Poland) as a gel matrix, and 0.1% (w/w) nitro blue tetrazolium chloride (NBT, Roth, Poland) as a radiation–sensitive compound. Each of the components was mixed with water using a magnetic stirrer with a temperature control system (IKA Works, Germany) as follows: (i) 4% w/w PVA at 85–90 °C (5 hours), (ii) 4% w/w borax at 60 °C (30 minutes), and (iii) NBT (0.1% of the final weight of the gel) at room temperature (15 minutes). Then, the PVA aqueous solution was mixed with the NBT aqueous solution and poured in layers, alternating with the borax aqueous solution into a beaker. The final solution was covered with Parafilm® and stored at room temperature (23 °C) for 48 hours before irradiation to form a gel. Then, the gel pieces were inserted into the low-density polyethylene string bag foils (LDPE, 40 mm×60 mm or 70 mm×100 mm, thickness 30±0.2 μm, Emerson, Poland) to maintain their flexible form. The samples in LDPE foils were placed on a steel table levelled with spirit levels and pressed down with a steel top foot. Between the stage and the top foot, four spacers of a certain height (5.2±0.1 mm) were used to keep the thickness of the samples constant.

2.2 Irradiation
The NBT–PVA gels were irradiated using a TrueBeam medical accelerator (Varian Medical Systems, USA; parameters: a photon beam of 6 MV, a monitor unit rate of 600 MU/min, field size 20×20 cm^2, dose range 0–50 Gy). The samples were immersed in a water phantom (W-1-DD1-2, GeVero Co., Poland) filled with distilled water (~21 °C) at a depth of 5 cm, SSD 95 cm.

2.3 Reflectance measurements
The reflectance spectra of the NBT–PVA gel samples were measured using a light reflectance instrument (Spectraflash 300, DataColor, Switzerland; parameters: illuminant D65, xenon lamp, angle 10°, a step of 10 nm, measurement error of 0.1%). The samples were measured 1 h after irradiation in the wavelength range of 400–700 nm. Based on reflectance measurements, the characteristic parameters of NBT–PVA gel were determined, such as threshold dose, dose sensitivity, as well as linear and dynamic dose ranges.

2.4 Spatial stability measurements
The spatial stability of the NBT–PVA gel was assessed after partial irradiation of a gel sample with 5000 MU of 6 MV FFF (flattening filter-free) photon beam (TrueBeam, Varian, USA); star-shot four fields with different collimator angle setup (0, 90, 150, and 240°). The field set by the jaws was 1×20 cm^2. The sample was scanned before irradiation and 1 h, 2 h, and 17 h after irradiation (HP Scanjet G3010 scanner, 300 dpi). The pictures obtained were split into RGB colour model channels and filtered (Mean filter, Kernel size: 11, Kernel unit: mm, mode of calculations: 2D) using the polyGeVero-CT software package (GeVero Co., Poland).

2.5. CT measurements
The NBT-PVA sample in syringes (50 ml, Webest, Medica, China) was measured with a Somatom Sensation Open CT scanner (Siemens AG Medical Solutions, Germany). It was placed in an electron density phantom (model 062, CIRS Norfolk, VA, USA) with a muscle-imitating insert. The scanning parameters were as follows: 120 kV, eff. mAs 115, pitch 0.5, slice thickness 1.5 mm, field-of-view (FOV) 500 mm. The B31f reconstruction algorithm was used (the spacing between images was 1.5 mm).

3. Results and discussion

The NBT–PVA gels are manufactured by the cross-linking of PVA chains and borate ions, which are formed as a result of the dissociation of borax in water [4]. After preparation, the gels are transparent and yellow in colour due to the addition of NBT dye. This colour changes to purple-

Journal of Physics: Conference Series **2799** (2024) 012007 doi:10.1088/1742-6596/2799/1/012007

brown under exposure to ionising radiation, which is related to the conversion of NBT salt into the formazan form. The change is more intense with a higher absorbed dose (Figure 1).

Figure 1. Photographs of NBT–PVA gels unirradiated and irradiated with different doses of radiation.

The changes occurring in NBT–PVA gels after irradiation with different doses of ionising radiation were registered using reflectance spectrophotometry. A decrease in light reflectance was observed with an increase in the absorbed dose of ionising radiation (Figure 2a). Based on these spectra, the dose-response of NBT–PVA gels was plotted, expressed as a relation between the reflectance at 540 nm (the wavelength of minimum reflectance) and the absorbed dose (Figure 2b). It allowed for the determination of the basic parameters of the dosimeter, such as dynamic range (from less than 2 to more than 50 Gy), linear range (from <2 Gy to 30 Gy), threshold dose (<2 Gy), dose sensitivity (0.90±0.07 Gy^{-1}), and intercept (48.25±1.04).

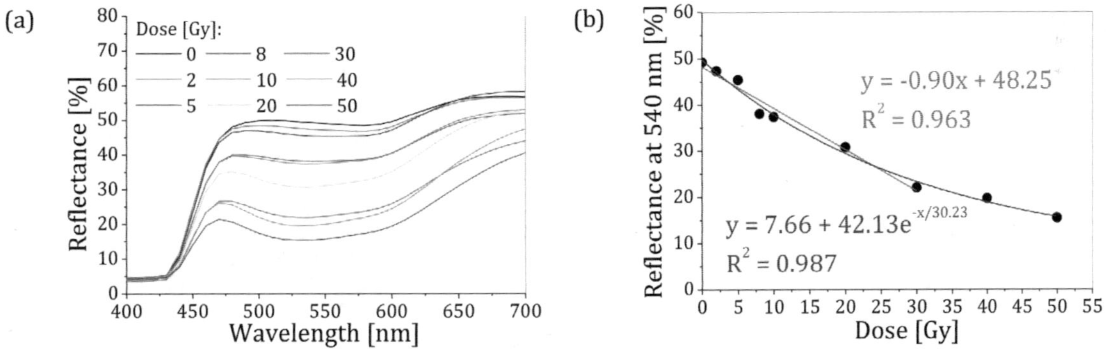

Figure 2. The reflectance spectra of the NBT–PVA gel irradiated in the dose range of 0–50 Gy (a), and the dose-response of this composition at a specific wavelength with fitted curves for dynamic (blue exponential curve) and linear (red linear curve) dose ranges (b). Each spectrum (a) and measurement point (b) is the average of five measurements calculated by the measuring device with an error of 0.1%.

The spatial stability of the NBT–PVA gel was examined to assess the possibility of performing dose distribution measurements. For this purpose, the gel was irradiated in the shape of three lines intersecting at one point. The raw sample image was decomposed into RGB (red-green-blue) channels using the polyGeVero-CT software package. The green channel was the most sensitive to changes occurring in the sample after irradiation (Figure 3a), and for this channel, the profile was plotted, passing from the unirradiated through the irradiated to the unirradiated part of the sample, which is marked in the figure with a red line (Figure 3b). Profiles were recorded 1, 2, and 17 h after irradiation. No diffusion of the irradiated part of the sample to the unirradiated part is observed during this time. The shape of the profile does not change, and there are small differences between the signal intensity of the unirradiated and irradiated parts of the sample measured at different times (12% between 1 and 17 h after irradiation, assuming the value of the unirradiated part at a distance of 0 mm and the irradiated part at 20 mm, Figure 3c). These differences and the shifts in the sample profiles result from the presence of air bubbles in the gel, which cause a decrease in signal intensity (dark fields, Figure

3b). In further work, steps should be taken to remove bubbles from the gel, e.g., using a vacuum chamber.

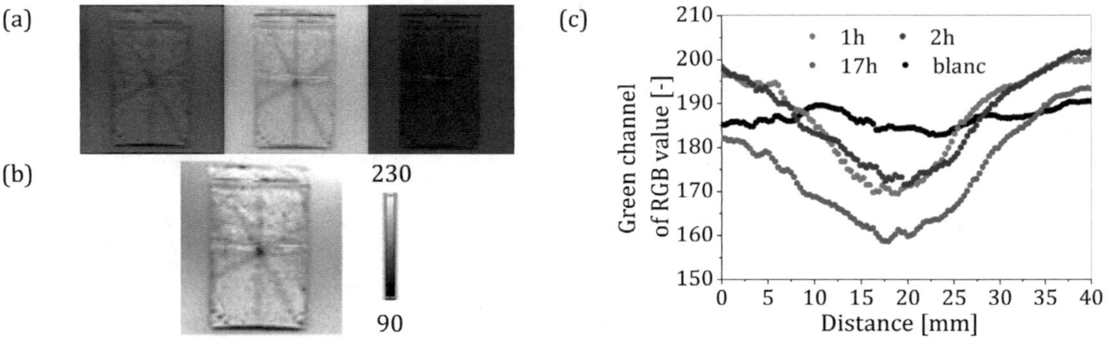

Figure 3. The NBT–PVA gel in the LDPE foil irradiated partially and decomposed into RGB channels (a). An image of the green channel (grayscale) with the red line indicates the position of profiles (b). Profiles for a blank sample (before irradiation) and samples measured over time after irradiation (c).

Irradiated gels become discoloured over time, as shown in Figure 4 for selected doses of ionising radiation. After 14 days after irradiation, complete discolouration of samples irradiated with lower radiation doses (2–10 Gy) is observed. Samples irradiated with higher doses (20–50 Gy) require longer time to completely discolour (they are under evaluation). After discolouration, the samples react again to ionising radiation, which indicates the possibility of reusing them.

Figure 4. Photographs of samples irradiated with a dose of 10 Gy (a) and 30 Gy (b), 1 h (left side) and 14 days (right side) after irradiation.

The HU value of the NBT-PVA gel equals 13 HU and is close to water (0 HU) and muscles (43 HU), which proves it is tissue-equivalent. The gel consists of 99.1% water and biocompatible poly(vinyl alcohol). However, due to the potential toxicity of borax and NBT to the skin, it is recommended to place the gel in the foil. The NBT-PVA gel is flexible and adapts to the curves of the body. Further research is necessary to determine the mechanical properties of the gel.

4. Conclusions

The new proposed radiochromic NBT–PVA gel shows potential for use as a reusable dosimeter in radiotherapy. Simultaneously, its flexibility and tissue-equivalence allow it to be used as a bolus, adhering closely to the curves of the body. The chemical composition and manufacturing procedure of the gel require further optimisation to remove air bubbles and obtain appropriate mechanical properties.

References

[1] Lu Y et al., 2021, International Journal of Bioprinting, 7, 27-42.
[2] Wang X et al., 2021, Frontiers in Oncology, 11, 698773.
[3] Dahn H M et al., 2021, Hematology, 163, 103391.
[4] Mahjoub H F et al., 2019, J. Mol. Liq., 291, 111272.

Journal of Physics: Conference Series **2799** (2024) 012008 doi:10.1088/1742-6596/2799/1/012008

Sorbitol to reduce Fe diffusion in a Fricke gel dosimeter and enhance its resistance to elongation

Michał Piotrowski[1], Piotr Maras[2] and Marek Kozicki[1,3*]

[1] Department of Mechanical Engineering, Informatics and Chemistry of Polymer Materials, Faculty of Materials Technologies and Textile Design, Lodz University of Technology, 90-543 Lodz, Poland
[2] Department of Radiotherapy Planning, Copernicus Hospital, 93-513 Lodz, Poland
[3]GeVero Co., Lodz, Poland

*E-mail: marek.kozicki@p.lodz.pl

Abstract. This work presents a 2D Fricke dosimeter with xylenol orange (XO), gelatine matrix and addition of sorbitol. The dosimeter was irradiated with various monitor units ranging from 500 to 10000 MU (3.6-72 Gy). The addition of sorbitol significantly reduced the diffusion coefficient of Fe^{3+} ions and improved the mechanical properties of the gel, while the chemical stability of the dosimeter decreased.

1. Introduction

Over the years, 3D dosimeters, both polymer and radiochromic [1-2] have been developed for testing irradiation instruments and verifying the cancer treatment plans. Recently, the use of 3D dosimeters has been proposed as a tool for examining accelerator parameters, such as the radiation isocenter [3]. The radiation isocenter can be determined by using both 2D flat dosimeters [4] and 3D dosimeters [5-7].

In this work, a 2D Fricke-XO-Gelatine dosimeter with sorbitol addition is presented. Attempts have already been made to add sucrose to the Fricke dosimeter [8], which has a similar impact on the properties of gelatine as sorbitol [9], resulting in a significant reduction in the Fe^{3+} diffusion coefficient. The addition of sugars also improves the mechanical properties of gelatine [9], making Fricke-XO-Gelatine with sorbitol a potential flexible dosimeter, bolus and tool to determine the coincidence of the radiation and mechanical isocentres of a medical accelerator. Similar flexible dosimeters have been previously described, such as LMG-silicone and FlexyDos3D [10-12].

2. Materials and methods

2.1 Dosimeter preparation

To prepare the Fricke-XO-Gelatine dosimeter, 6% gelatine (Sigma, USA), 50 mM sulphuric acid (Chempur, Poland), 0.5 mM ferrous ammonium sulphate (FAS, Chempur, Poland), 0.165 mM xylenol orange (Sigma-Aldrich, Germany) and 23% sorbitol (Biomus, Poland) were used. Initially, gelatine was added to 65% of the final weight of water at 55°C (double distilled water was used). After the gelatine was completely dissolved, the temperature was lowered to 35°C. At the same time, sorbitol was added to 30% of the final water weight at 45°C, and then, after complete dissolution, the temperature was lowered to 35°C. Sulphuric acid, FAS and XO were dissolved in the remaining water mass, then added to the gelatine solution and mixed gently. Finally, the sorbitol solution was added to the gelatine with Fricke components and stirred gently at 35°C.

Journal of Physics: Conference Series **2799** (2024) 012008 doi:10.1088/1742-6596/2799/1/012008

2.2 Sample preparation

The dosimetric sample was prepared by pouring the dosimetric solution into a square frame made of poly(methyl methacrylate) (PMMA) with internal dimensions of 10 cm × 10 cm × 0.5 cm. The frame was placed on a sheet of rigid foil and then placed on a levelled stainless-steel plate. After pouring the solution, the frame was covered with rigid foil, and then two more stainless steel plates were put on top of the sample to remove excess solution. After 15 minutes, the plates were removed, and the frame covered with foil was placed in the refrigerator for another 60 minutes.

2.3 Irradiation

The sample was irradiated using a medical accelerator (TrueBeam, Varian, USA). The dosimeter was irradiated with the following monitor units (MUs): 500, 1000, 1500, 2000, 2500, 5000, 7500, and 10000 MU, corresponding to 3.6, 7.2, 10.8, 14.4, 18, 36, 54, 72 Gy, respectively. The remaining irradiation parameters were as follows: X-rays, 10 MV FFF (flattening filter free), monitor unit rate of 2400MU/min, the leaf gap of high-definition multi-leaf collimator (HD MLC) 0.5 cm, jaws size X: 4 cm, Y: 3 cm, gantry and collimator set to 0°, SSD 100 cm. The sample was placed on five slabs (the thickness of each slab was 1 cm) of the SP34 RW3 phantom (IBA, Germany), perpendicular to the central axis of the radiation beam. The sample was covered with two more slabs (1 cm and 0.5 cm thick).

2.4 Readout and data processing

Irradiated samples were scanned with an HP Scanjet G3010 flat-bed scanner under the following conditions: 75 dpi, brighten/darken option: 0 (brighten), -52 (shadows), 0 (intermediate shades), sharpen the image: none, automatic correction of colour: none, colour adjustment: none (100% saturation of colour). Scanning was performed 47 min before and 32, 80, 200, 240 and 1200 min after irradiation. Image processing was performed with the polyGeVero-CT software package (GeVero Co., Lodz, Poland).

3. Results and discussion

Fricke-XO-Gelatine dosimeter with sorbitol was exposed to different monitor units and scanned for the next 20 hours using an HP Scanjet G3010 flatbed scanner. The images of the scanned dosimeter before and after irradiation are presented in Figure 1. Just after preparation, the sample was dark orange, then, within the next two hours, it darkened to brownish, and three hours after preparation it started turning blue, which indicates a high autooxidation rate of the Fe^{2+} ions. A similar effect was reported with the addition of sucrose and glycerol [8]. The low chemical stability of ferrous ions is probably due to too high concentrations of gelatine and sorbitol. Another disadvantage of using high concentrations of gelatine and sorbitol is the presence of air bubbles in the sample due to the high viscosity of the dosimeter solution.

The image of the sample scanned 32 minutes after irradiation was resolved into red, green and blue (RGB) channels. For each channel, profiles across the irradiated areas (Figure 2 A and B) were determined to find which colour changes the most after irradiation. It was found that the most significant change occurs for the red channel, which results in the use of this channel in further data analysis. The linear response of the dosimeter is in the range of 3.6 to 14.4 Gy (Figure 2 C).

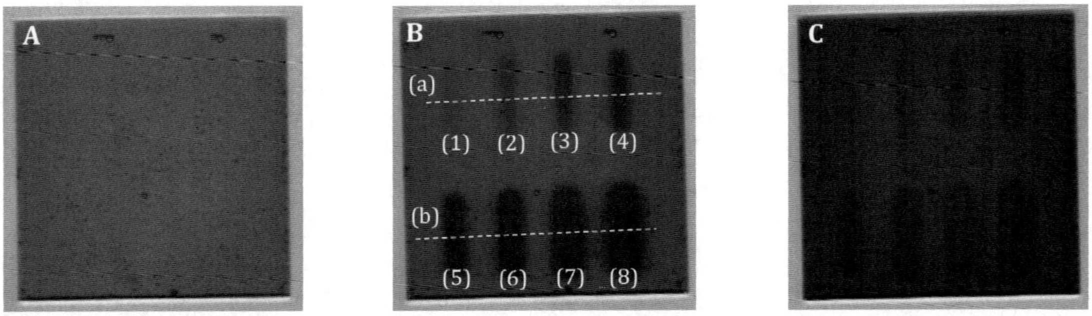

Figure 1 Response of the Fricke-XO-Gelatine dosimeter with sorbitol to irradiation. A: Sample of the dosimeter just after preparation, B and C: the same sample 32 and 240 minutes, respectively, after irradiation (time between preparation and irradiation was 47 minutes); the sample scanned with an HP Scanjet G3010 flat-bed scanner. The areas were irradiated with 500 (1), 1000 (2), 1500 (3), 2000 (4), 2500 (5), 5000 (6), 7500 (7), and 10000 (8) MUs.

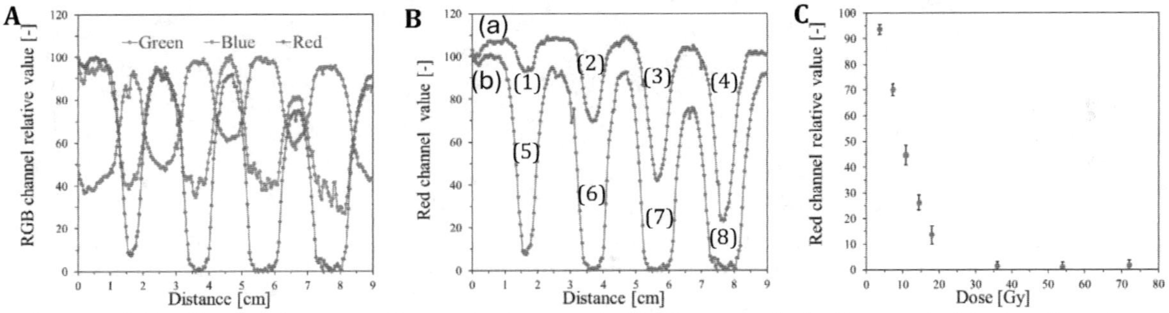

Figure 2 A: Profiles for the green, blue and red channels (RGB) across the bottom part of the irradiated areas as indicated in Figure 1C by the dashed line (b). B: Profiles of the irradiated areas for the red channel. Profiles (a) in B correspond to the top part (dashed line (a) in Figure 1B), while profiles (b) in B correspond to the bottom part (dashed line (b) in Figure 1B). Numbers 1-8 correspond to the irradiated areas as presented in Figure 1B. C: Red channel relative value versus dose based on profiles showed in B.

The profiles across the irradiated areas determined 32, 80 and 240 minutes after irradiation are presented in Figure 3. A change in the value of the red channel is visible for each monitor unit value used. The maximum value of the red channel for the irradiated areas does not change within four hours after irradiation, indicating a stable radiochromic response of the dosimeter with sorbitol which is an advantage over the dosimeter with sucrose [8], which changes the colour of irradiated area even for several hours after irradiation. For irradiations corresponding to (6)-(8) (Figure 2), the saturation of colour occurred. The width of the profiles almost does not change within 240 minutes after irradiation. For example, the calculated values of the profile width at half its height for the area irradiated with 1500 MU were 4.90 mm and 4.94 mm after 32 and 240 minutes after irradiation, respectively. The uncertainty of the obtained results is 0.34 mm. It shows a potentially low diffusion of ferric ions, however, the diffusion coefficient should be determined in the future.

The mechanical properties of the Fricke-XO-Gelatine dosimeter with the addition of sorbitol are shown in Figure 4. The sorbitol contained in the dosimeter significantly improves the dosimeter's tensile strength. This feature can be further used for manufacturing both deformable and bolus dosimeters.

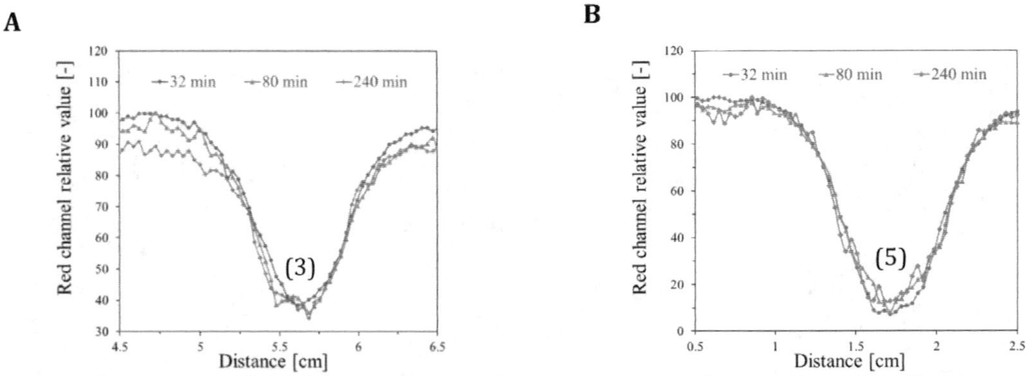

Figure 3 Example profiles of the irradiated areas presented in Figure 1B and C. Numbers 3 and 5 correspond to irradiated areas as presented in Figure 1B.

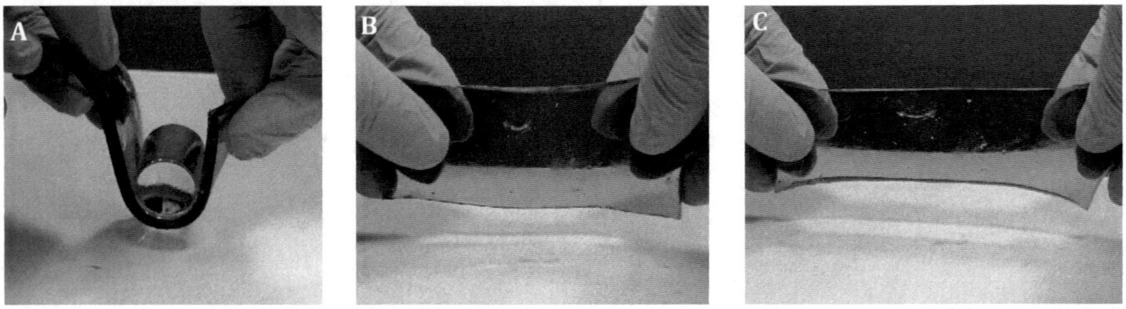

Figure 4 A: dosimeter sample with a 100 g stainless steel weight placed in the center of the sample held in the hands above the table surface. B and C: Hand-held sample before (B) and after (C) manual stretching.

4. Conclusions

This work presents a Fricke-XO-Gelatine dosimeter with a high gelatine concentration and the addition of sorbitol. The dosimeter exhibits reduced chemical stability, which is most likely due to too high concentration of gelatine and sorbitol, and the high viscosity of the dosimeter solution may cause the presence of air bubbles in the gel. The results indicate that the diffusion coefficient of Fe^{3+} ions is reduced, but further research is required. The factor limiting the use of this dosimeter is the rapid autoxidation of ferrous ions. Additional studies require optimisation of gelatine and sorbitol concentrations to achieve greater stability without increasing the diffusion coefficient. To evaluate the dosimeter's suitability as a bolus or deformable dosimeter, mechanical properties should be determined by performing additional tensile strength and fatigue tests.

References

[1] Baldock C *et al*, 2010, *Phys. Med. Biol.*, **55**, R1-R63
[2] De Deene Y, 2022, *Gels*, **8**, 599
[3] Dorsh S *et al*, 2018, *Phys. Med. Biol.*, **63**, 11NT02
[4] Gonzales A *et al*, 2004, *Med. Phys.*, **31**, 1489-1493
[5] Pant K *et al*, 2020, *Med. Phys.*, **47**, 927-936
[6] Maras P and Kozicki M, 2022, *Materials*, **15**, 6807
[7] Kozicki M and Maras P, 2023, *Measurement*, **221**, 113585
[8] Jordan K and Sekimoto M, 2010, *J. Phys.: Conf. Ser.*, **250**, 012048
[9] Dai H *et al*, 2020, *Food Hydrocolloids*, **101**, 105557
[10] Høye E M *et al*, 2015, J. *Phys.: Conf. Ser.*, **573**, 012067
[11] Jensen S V *et al*, 2020, *Phys. Imaging Radiat. Oncol.*, **16**, 134
[12] De Deene Y *et a*l, 2015, *Phys. Med. Biol.*, **60**, 1543

Journal of Physics: Conference Series **2799** (2024) 012009 doi:10.1088/1742-6596/2799/1/012009

A characterization of LCV micelle gel dosimeters for boron neutron capture therapy

Ryosuke Narita[1], Shin-ichiro Hayashi[2] and Yoshinori Sakurai[3]

[1]Graduate School of Engineering, Kyoto University, Kyoto 615-8540, Japan
[2]Hiroshima International University, Hiroshima 739-2695, Japan
[3]Institute for Integrated Radiation and Nuclear Science, Kyoto University, Osaka 590-0494, Japan

Email: sakurai.yoshinori.8n@kyoto-u.ac.jp

Abstract. As a method for three-dimensional quality assurance in boron neutron capture therapy (BNCT), micelle gel dosimeters are focused on. We are investigating dose-component discrimination method by using multiple gel dosimeters which have each different radiation quality specificity. In this study, the characteristics of each gel dosimeter for neutrons and gamma-rays were evaluated, and a dose-component discrimination method was suggested.

1. Introduction

In boron neutron capture therapy (BNCT), the dose estimation is based on treatment planning simulation, and the quality assurance (QA) is performed by measuring actual doses to ensure treatment accuracy [1]. Generally, the measurement methods using activation foil and thermo-luminescent dosimeter (TLD) are recommended. However, these methods are only zero or one-dimensional dose estimations, and more complex two or three-dimensional (3D) estimations are expected. The dose in BNCT is estimated from the four dose components, such as boron, nitrogen, hydrogen and gamma-ray doses. Therefore, the discrimination estimation of these dose components is necessary. Gel dosimeter is one of the dosimeters for the estimation of 3D dose distribution, and its elemental composition can be modified relatively easily. Furthermore, it can be sensitive to specific dose components depending on the gel composition. So, it is expected to be applied to the estimation in BNCT [2].

Gel dosimeters such as Fricke gel dosimeter [3], polymer gel dosimeter [4], and PVA-GTA-I gel dosimeter have been investigated for QA in BNCT. These suggest the method for improving specific radiation quality sensitivity for neutrons [5, 6]. Applying these methods, we have been investigated basic characteristics of micelle gel dosimeters for gamma-rays for dose-component discrimination by our previous studies [7]. In this study, the neutron and gamma-ray characteristics of micelle gel dosimeters improved for QA in BNCT were evaluated by mainly experimental studies, and a dose-component discrimination method using gel dosimeters was suggested.

2. Materials and methods

2.1. Preparation of gel dosimeters

Leucocrystal violet (LCV) was selected as dye of micelle gel dosimeters, because it is extremely clear before irradiation. The micelle gel dosimeters were prepared based on 1 mM of LCV in the 96% water and 4% gelatin according to the method of Babic et al. [8]. For dose rate independency and temporal coloration reduction, the type and concentration of surfactant and sensitizer were changed such as 6 mM of cetyltrimethylammonium bromide (CTAB), 30 mM of trichloroacetic acid (TCAA), and 100 mM of 2-2-2 trichloro ethanol (TCE). In addition, 50 mM of boric acid or 2000 mM of urea was added for the estimation of boron or nitrogen dose components, referring to Gambarini et al. [5]. Furthermore, a dosimeter in which light water was replaced with heavy water was also prepared to estimate hydrogen dose component [6], as shown in Table 1. The amount of additives was based on our previous study. To measure the light absorbance, the gel dosimeters were dispensed into 1×1×4.5 cm^3 polystyrene disposable cells for spectrophotometers and stored in a 4°C incubator.

Table 1. The materials and composition modified for each gel dosimeter

Materials	Standard	Heavy-water-based	Boric-acid-added	Urea-added
Water [wt%]	96	0	96	96
Heavy water [wt%]	0	96	0	0
Boric-acid [mM]	0	0	50	0
Urea [mM]	0	0	0	2000

* wt%: weight percentage. M: molar (mol/1000 g H_2O).

2.2. Gamma-ray and neutron irradiation

The gamma-ray irradiation was performed using the Co-60 gamma-ray irradiation system at Institute for Integrated Radiation and Nuclear Science, Kyoto University (KURNS). For dose-response evaluation, the irradiation was performed at doses of 0, 4, 8, 12, 16 and 20 Gy. Absorbed dose was measured with TLDs (UD-170LS, Panasonic Co.).

Heavy Water Neutron Irradiation Facility installed in Kyoto University Reactor (KUR-HWNIF) was used as the neutron irradiation field [9]. In KUR-HWNIF, the irradiation modes with different neutron energy spectra can be selected according to the application. In this study, the irradiations were performed using the standard mixed neutron irradiation mode at the reactor power of 5 MW for 3 hours, and the standard thermal neutron irradiation mode at 1 MW for 3 hours. The gold-foil activation method and TLDs were used for the estimation of neutron and gamma-ray doses, respectively. During the irradiation, the four types of gel dosimeters prepared in *Section 2.1.* were placed, as three of each type were bundled together. The differences in spatial dose distribution were corrected by the measured values of the gold foils and TLDs, which were attached to each gel dosimeter.

Absorbance spectra for 580 nm to 650 nm were obtained using a spectrophotometer (V730, Japan Spectroscope Co.). The absorbance at 600 nm, where the peak exists, was evaluated.

2.3. Simulation and dose-component discrimination

To evaluate the dependency for linear energy transfer (LET) from the results in *Section 2.2.*, the LET distribution inside the gel dosimeter was calculated using a Monte Carlo simulation code, PHITS [10], and the dose-averaged LET was evaluated. The geometry for *Section 2.2.* was modelled. The nominal neutron energy spectra of KUR-HWNIF were input in the neutron source, and the normalization was performed based on the neutron energy spectrum near the irradiation aperture.

For the dose-component discrimination, a 4×4 matrix equation as shown in equation (1) was considered. The factors "*k*" were calculated from the gel dosimeter response "*A*" to the absorbed dose "*D*" of each dose component. The neutron and gamma-ray doses "*D*" were estimated from the gold-foil activation method and TLDs, and the responses "*A*" were obtained from the measurement by spectrophotometer, as described in *Section 2.2.*.

$$\begin{pmatrix} k_{H,s} & k_{N,s} & k_{B,s} & k_{G,s} \\ k_{H,b} & k_{N,b} & k_{B,b} & k_{G,b} \\ k_{H,u} & k_{N,u} & k_{B,u} & k_{G,u} \\ k_{H,h} & k_{N,h} & k_{B,h} & k_{G,h} \end{pmatrix} \begin{pmatrix} D_H \\ D_N \\ D_B \\ D_G \end{pmatrix} = \begin{pmatrix} A_s \\ A_b \\ A_u \\ A_h \end{pmatrix} \tag{1}$$

D_H, D_N, D_B, D_G: Doses for hydrogen, nitrogen, boron and gamma-ray components
A_s, A_b, A_u, A_h: Absorbances for standard, boric-acid-added, urea-added and heavy-water-based gel dosimeters
$k_{H,s} \cdots k_{G,h}$: Parameters of each gel dosimeter for each dose component

3. Results and discussion

3.1. Neutron and gamma-ray characterization

As shown in Figure 1, the dose response for gamma-rays showed that only the urea-added gel dosimeter was approximately twice as large as other gel dosimeters. Note that the error bars indicate the 2σ of the residual standard error. The cause is not clear but is may depend on the solvent polarity and pH. On the other hand, as shown in Figure 2, the dose response for neutrons showed that the response of boric-acid-added dosimeter was almost one tenth of that of the standard dosimeter, but otherwise the responses were almost the same as the gamma-ray responses. Then, the error shows minimum and maximum values for the three dosimeters. It was considered that the boric-acid-added dosimeter has a high LET of secondary particles, so the neutron dose response has decreased.

Therefore, the LET dependence was evaluated from the convolution of experiments and simulations, as shown in Figure 3. Here, the "dose response rate" is the dose response of the dose component based on the experiments and equation (1) (Derivation formula was detailed in *Section 3.2.*.) divided by the gamma-ray dose response based on the gamma-ray experiment. LET dependence distribution showed that the response rate decreases at the high dose-averaged LET above 100 keV/μm. It is considered from the mechanism of coloration that the radical recombination amount increases with radical density increase, and the dosimeter response was decreased above 100 keV/μm.

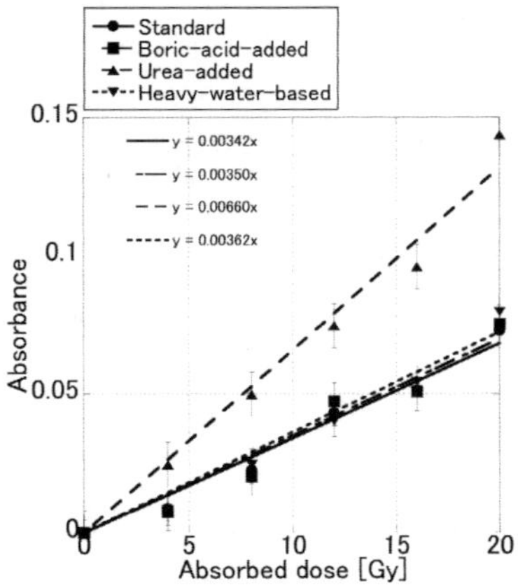

Figure 1. Dose response for gamma-rays

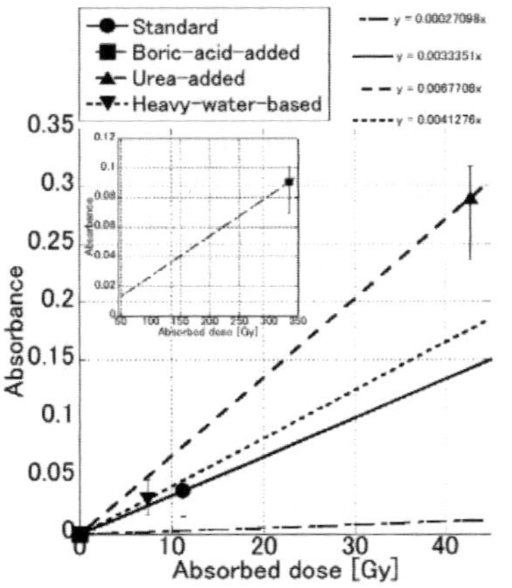

Figure 2. Dose response for neutrons (mixed mode)

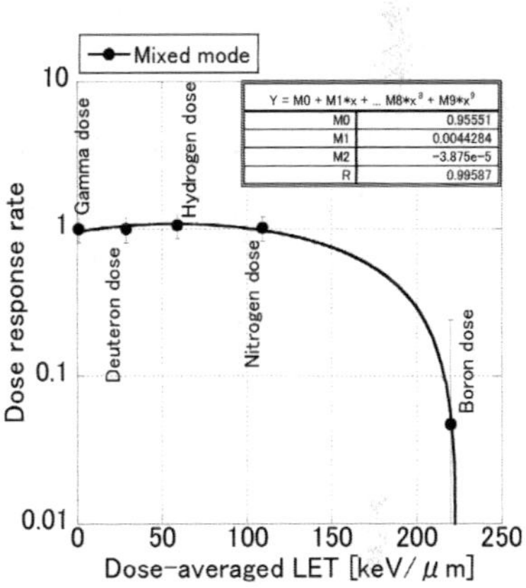

Figure 3. LET dependence

3.2. The 4×4 matrix factors for dose-component discrimination

The slope of each line in Figure 1 is the parameter of "k" for gamma-ray dose. Due to the influence of urea, the parameter "$k_{G,u}$" for urea-added dosimeter is larger than the other parameters. This factor of urea for dose response ratio is defined as "F_u". The parameters of the other dosimeters for gamma-ray dose are almost the same.

The parameters of dosimeters for boron dose except for that of boric-acid-added dosimeter are zero. Since the components except for boron-acid are the same, it is assumed that the standard dosimeter and boric-acid-added dosimeter have the same parameters for nitrogen dose and hydrogen dose. By taking the difference between the expanded equations from the equation (1) for standard dosimeter and boric-acid-added dosimeter, the parameter of boric-acid-added dosimeter for boron dose "$k_{B,b}$" can be evaluated as "$(A_b - A_s)/D_B$".

The parameter of urea-added dosimeter for nitrogen dose "$k_{N,u}$" can be evaluated from the expanded equations for standard dosimeter and urea-added dosimeter. These dosimeters have greatly different concentrations of nitrogen and slightly different concentrations of hydrogen, so it is necessary to make corrections not only for the factor "F_u" but also for the concentrations of hydrogen and nitrogen. As the concentration ratios of hydrogen and nitrogen for urea-added dosimeter to standard dosimeter are defined as "$C_{H,u}$" and "$C_{N,u}$", the parameters "$k_{H,u}$" and "$k_{N,u}$" are expressed as "$F_u \times C_{H,u} \times k_{H,s}$" and "$F_u \times C_{N,u} \times k_{N,s}$", respectively.

The parameter of standard dosimeter for hydrogen dose can be evaluated from the expanded equations for standard dosimeter and heavy-water-based dosimeter. As the ratio of effective kerma factors for hydrogen to deuterium is defined as "$K_{H/D}$" and the ratio of number density for hydrogen in standard dosimeter to deuterium in heavy-water-based dosimeter is defined as "$R_{H/D}$", it is assumed that "$k_{H,s}$" is approximately equal to "$K_{H/D} \times R_{H/D} \times k_{H,h}$".

The relationship between the parameters is summarized in equation (2).

$$\begin{gathered} k_{H,s} \fallingdotseq k_{H,b}, \quad k_{H,u} = F_u \times C_{H,u} \times k_{H,s}, \quad k_{H,s} \fallingdotseq K_{H/D} \times R_{H/D} \times k_{H,h} \\ k_{H,s} = \{(A_s - A_h) \times (K_{H/D})/(K_{H/D} - 1/R_{H/D})\}/D_H \\ k_{N,s} \fallingdotseq k_{N,b} \fallingdotseq k_{N,h}, \quad k_{N,u} = F_u \times C_{N,u} \times k_{N,s} \\ k_{N,u} = \{(A_u - F_u \times A_s) - F_u \times (C_{H,u} - 1) \times k_{H,s} \times D_H\}/\{F_u \times (C_{N,u} - 1) \times D_N\} \\ k_{B,s} = k_{B,u} = k_{B,h} = 0, \quad k_{B,b} = (A_b - A_s)/D_B \\ k_{G,s} \fallingdotseq k_{G,b} \fallingdotseq k_{G,h} = k_{G,u}/F_u \end{gathered} \tag{2}$$

The 4×4 matrix of "k" for mixed mode was evaluated in equation (3). These factors were modified by $K_{H/D}$ for the thermal neutron irradiation mode and experimentally verified. In this verification, the light absorbances obtained for the gel dosimeters were compared with the light absorbances estimated based on the data measured using gold foils and TLDs. The light-absorbance values obtained using the both methods were almost the same, the difference of each was only 4.0~6.3 %, and it was included in the measurement errors of each measurement method (Approximately, ~0.0081 Abs. for the gel method, and ~15 % for the gold and TLD methods, for the standard errors.). Therefore, it was confirmed that the "k" factors in equation (3) was appropriate for dose-component discrimination.

$$k_{H,s} \cdots k_{G,h} = \begin{pmatrix} 0.0036 & 0.0003 & 0.0000 & 0.0034 \\ 0.0036 & 0.0003 & 0.0002 & 0.0034 \\ 0.0074 & 0.0067 & 0.0000 & 0.0068 \\ 0.0018 & 0.0003 & 0.0000 & 0.0034 \end{pmatrix} \tag{3}$$

4. Conclusion

The characterization of LCV micelle gel dosimeters was performed and dose response for neutrons and gamma-rays was evaluated. Using these results, the factors for dose-component discrimination method were determined and verified for different energy spectra, and the validity was confirmed.

5. Acknowledgement

This study was supported by JST SPRING (Grant No. JPMJSP2110).

References

[1] Liu Y H et al 2011 *Appl. Radiat. Isot.* **69**(12), 1897-1900
[2] Gambarini G et al 2004 *Appl. Radiat. Isot.* **61**(5), 759-763
[3] Gambarini G et al 1997 *Appl. Radiat. Isot.* **48**(10-12), 1477-1484
[4] Hayashi S et al 2015 *J. Phys.: Conf. Ser.* **573**, 012074
[5] Gambarini G et al 1994 *Nucl. Instrum. Meth. A* **353**(1-3), 406-410
[6] Gambarini G et al 2001 *IEEE Trans. Nucl. Sci.* **48**(3), 780-784
[7] Narita R & Sakurai Y 2023 *J. Phys: Conf. Ser.* **2630**, 012016
[8] Babic S et al 2009 *Phys. Med. Biol.* **54**(22), 6791-6808
[9] Sakurai Y & Kobayashi T 2000 *Nucl. Instr. Meth. A* **453**(3), 569-596
[10] Sato T et al 2024 *J. Nucl. Sci. Tech.* **61**, 127-135

Journal of Physics: Conference Series **2799** (2024) 012010 doi:10.1088/1742-6596/2799/1/012010

Simultaneous comparison of two independent dosimetry verification systems for VMAT irradiation in assessing 4D dose distribution

Aurimas Krauleidis[1*], Diana Adlienė[1]

[1] Kaunas University of Technology, Department of Physics, Studentu Str. 50, 51368 Kaunas, Lithuania

*E-mail: aurimas.krauleidis@ktu.edu

Abstract. Despite the rapid development of new technologies, 4D dose assessment related to the breathing movement of target organs during radiotherapy treatment is still a challenge. This study aimed to investigate the feasibility of polymer gel dosimeters to assess 4D dose distribution accuracy in VMAT radiotherapy treatment under dynamic conditions and to compare assessment results with the data obtained using a cylindrical water-equivalent phantom with a three-dimensional diode array (ArcCheck). For this purpose, several homemade polymer gels were filled in cylindrical containers and placed within the ArcCheck phantom. The entire system was mounted on rotating wheels and periodically moved by a connected robotic platform during irradiation. Generally, good agreement was found between measurement results of the 4D dose distribution obtained in the bulk of volume using homemade polymer gel dosimeters and diode detector system, indicating the potential to address the remaining small errors in further experiments. The latter implies the applicability of polymer gel dosimeters for the validation of high-dose stereotactic treatment plans under realistic 4D dose distribution conditions.

1. Introduction

Modern adaptive radiotherapy delivery techniques have become increasingly complex in recent years. The success of radiotherapy treatment of cancer patients depends on the proper planning and delivery of doses in each stage of the treatment. Even minor errors can compromise the effectiveness of the treatment. However, advanced delivery techniques are often associated with moving beams, small fields, steep dose gradients and the effect of tumour motion. The most challenging issue is the application of respiratory gating techniques related to the tumour movement due to the patient's breathing cycle [1]. The tumour can follow complex trajectories, and the breathing pattern may vary depending on the intensity, duration, regularity and starting position. Breathing related movements are contributing to the discrepancies between planned and delivered to the target dose distributions [2]. Application of respiratory gating technique aimed at reduction of dose delivery to the healthy tissues without compromising the established target coverage and shortening of treatment duration, its clinical implementation requires validation of the intended workflow using dynamic phantom experiments and 4D dose measurements [3].

Polymer gel dosimeters can serve as independent 4D detectors having desired characteristics, like equivalency to biological tissues, accuracy and precision and providing ability to resolve three-dimensional (3D) dose distributions in dynamic dose verification [4, 5]. Polymer gel dosimeters can be used as an introduction to new benchmarking techniques in the clinic, adding an extra dimension to ongoing pursuit improved patient safety.

Moreover, in most studies a fixed setup measurements are investigated and data cannot show the dynamic dose distribution. In this study nMAG and nPAG polymer gel dosimeters were coupled to simulated dynamic organ displacement caused by motion phantom with a diode array detector.

2. Materials and methods

In this study, the ArcCheck verification system (model 1220) was used to simulate the patient's body and perform reference dose distribution measurements. Independent verification system is represented by a cylindrical water-equivalent phantom with a three-dimensional array containing of 1386 diode detectors, arranged on a phantom according to a helical geometry with

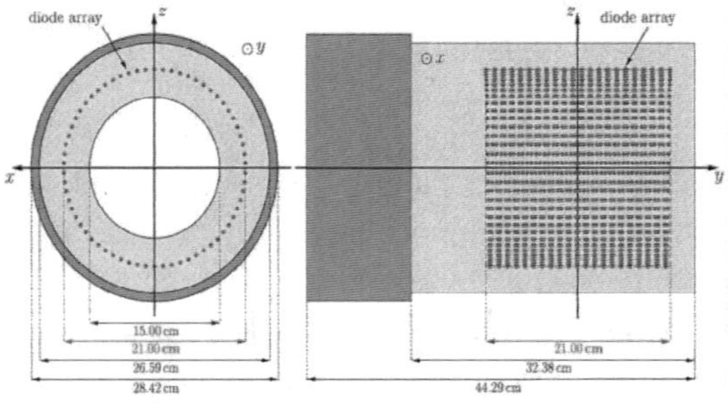

Figure 1. The geometry of the ArcCheck device.

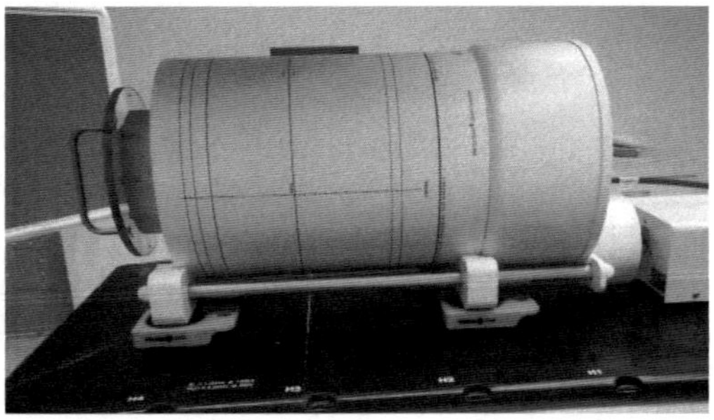

Figure 2. Experimental setup of the system.

10 mm spacing in between (Figure 1). The diameter and size of the diode array is 21 cm. Each diode in the array has a sensitive volume of 0.019 mm^3 and provides updated dose measurement results within 50 ms intervals. The array itself is situated at a water equivalent depth of 3.3 g/cm^2. The central cavity of the device is cylindrical, having diameter of 15 cm. The experimental setup of the system (Figure 2) shows the ArcCheck, gel vial (inside the ArcCheck), wheels and motion phantom.

Normoxic methacrylic acid polymer gel dosimeter (nMAG) and normoxic polyacrylamide polymer gel dosimeter (nPAG) was used for 3D dose measurements. Polymer gel dosimeter 's recipes were taken and fabrication procedure was implemented following information provided in previous publications [6, 7]. Prepared polymer gels were poured into cylindrical PMMA containers (H=100mm, Ø 150 mm) and tightly closed to avoid oxygen contamination. To perform 4D dose measurements 4 containers were filled with two different polymer gels and placed in the centre of the ArcCheck phantom. Four wheels were attached to the phantom to enable its movement. At least one container filled with corresponding polymer gel dosimeters was not irradiated and left for reference.

The simulation of target 's movement was performed using respiratory gating robot (Anzai Medical) connected to an ArcCheck phantom that was set in sinusoidal motion along longitudinal direction. Movement rate of 15 rpm with an amplitude of 20 mm was chosen. During irradiation activated movement resulted in a simultaneous dynamic signal recording process by detector

Journal of Physics: Conference Series **2799** (2024) 012010 doi:10.1088/1742-6596/2799/1/012010

array and polymer gel dosimeter. 2 containers filled with different polymer gels were irradiated during movement and another 2 containers without movement.

A real lung stereotactic treatment plan with a single dose of 24 Gy was created by treatment planning system (Eclipse version 15.6) of RapidArc treatment modality using calculation algorithm AAA (Anisotropic Analytical Algorithm) with a calculation resolution of 1.25 mm. A linear accelerator (TrueBeam, Varian) was used to deliver the dose to a moving phantom with the polymer gel dosimeters inside. 6 MV FFF photon beam with a maximum monitor unit rate of 1400 MU/min was used for irradiation. The irradiation plan was realized using four partial arcs with total 9474 MU's. A calibration curve was obtained with doses ranging from 0 to 36 Gy.

Irradiated polymer gel dosimeters were evaluated in MRI facility analysing polymerization induced alteration of the relaxation rate R2 of the transverse magnetization depending on the absorbed radiation dose. Irradiated polymer gel dosimeters were scanned with a 1.5 T MRI unit Magnetom Aera (Siemens Healthineers) using the following imaging parameters: TR = 3000 ms, 32 equidistant echoes, TE = 22.5 – 720 ms, resolution = $1 \times 1 \times 1$ mm^3. In parallel CT readouts of irradiated polymer gel dosimeters were performed using GE Discovery CT scanner. The CT scanning parameters were as follows: tube voltage - 140 kVp, tube current - 320 mAs, and slice thickness of 1 mm. Eclipse 15.6, IQ Works, ImageJ and Origin Pro software's were used for data processing and analysis. For CT and MRI data ROI was selected as a mean of area plot with the applied background subtraction. A dose profile was measured through the centre of the target. Image filtering was not applied. Additionally, special home-based algorithm developed using a MatLab R2021a programming platform and SNC patient software was applied to create dose distributions in the form of dose maps.

3. Results and discussion

The total effect of motion on the dose delivery to the target and surrounding tissues was evaluated, i.e. the 3D dose distribution difference between measurement results obtained using polymer gel dosimeters and ArcCheck detector during phantom motion. The blurring effect was identified when investigating dose distribution in polymer gel dosimeters irradiated under a moving system setup (Figure 3 and Figure 4).

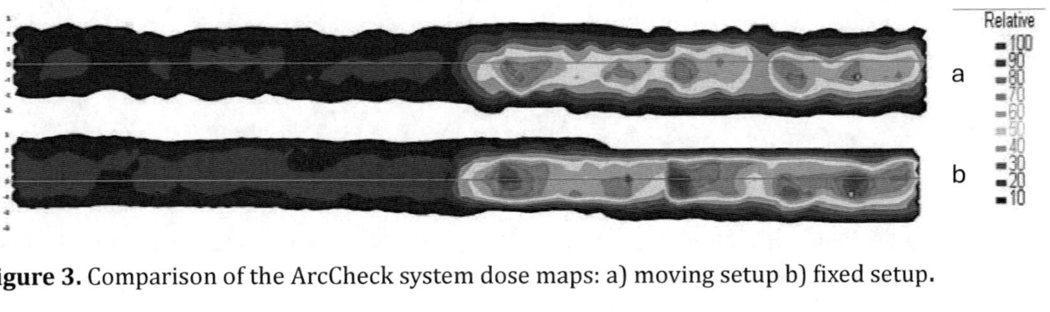

Figure 3. Comparison of the ArcCheck system dose maps: a) moving setup b) fixed setup.

It can be noticed that polymer gel dosimeters are less sensitive at lower dose gradients. However, dose values obtained using polymer gel dosimeters and ArcCheck were in good agreement within homogeneous areas. The average γ passing rates applying 3%/3 mm or 3%/2 mm criteria and the threshold of 10% were 99.9 % for ArcCheck system in the fixed setup compared to the target plan calculated by the treatment planning system, so we can consider the ArcCheck results as reference data.

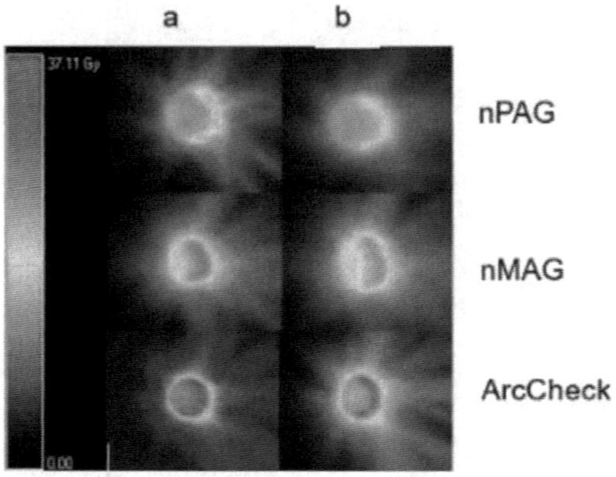

Figure 4. Dose distribution maps in the target a) fixed system setup and b) moving system setup.

For nMAG polymer gel dosimeter γ passing rates were greater than 90 % and for nPAG polymer gel dosimeter quite low, around 38 % respectively. For moving system setup γ passing rate applying 3%/3 mm was 51,3 %, 45,3 % and 27 % respectively (Figure 5 and Figure 6). The calculated percentage dose difference (DD, %) for the planning target volume (PTV) mean dose and maximum dose reached up to 32 %.

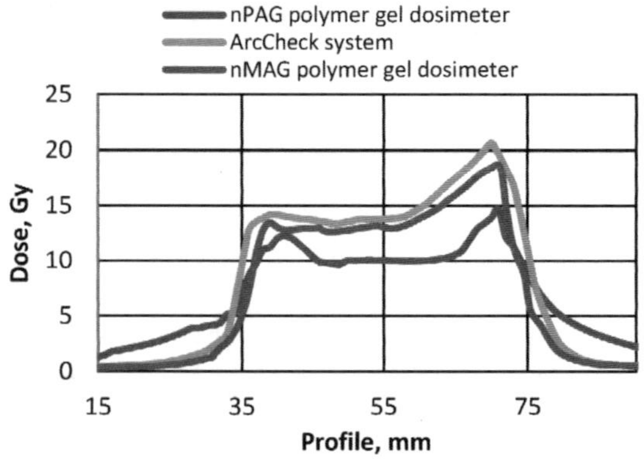

Figure 5. Comparison of the target dose profiles obtained using polymer gel dosimeters and the ArcCheck system in moving setup.

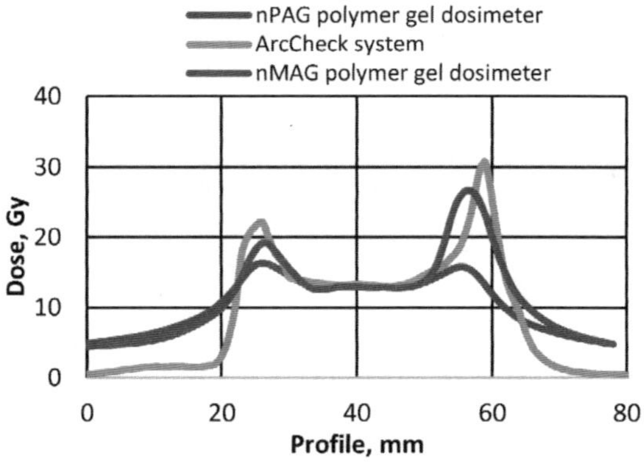

Figure 6. Comparison of the target dose profiles obtained using polymer gel dosimeters and the ArcCheck system in a fixed setup.

4. Conclusion

Investigated polymer gel dosimeters indicated their applicability for dose measurements during target movement and for dose verification of dynamic RT involving gated dose delivery. The combination of moving diode phantom with 3D polymer gel dosimeters might be a valuable tool for validating adaptive radiotherapy.

References

[1] Netherton T et al 2018 *Med. Phys.* **45** 2369-76
[2] Edvardsson A et al 2018 *Phys. Med. Biol.* **63** 085012
[3] Bäck S Å J et al 2019 *J. Phys.: Conf. Ser.* **1305** 012049
[4] Krauleidis A et al 2024 *Nucl. Instrum. Meth. B.* **550**, 165316
[5] Karlsson A et al 2007 *Phys. Med. Biol.* **52**(15) 4697-4706
[6] Krauleidis A et al 2022 *Gels* **8** 629
[7] Deene Y De et al 2006 *Phys. Med. Biol.* **51** (3) 653-73

Journal of Physics: Conference Series **2799** (2024) 012011 doi:10.1088/1742-6596/2799/1/012011

Clinical Applications and Monte Carlo Validation of a fully integrated 3D dosimetry system

Kawtar Lakrad[1,2*], Benjamin Quinn[3], Mark Oldham[2] and Justus Adamson[2]

[1] Department of Physics, Hassan II University, Casablanca, Morocco
[2] Department of Radiation Oncology, Duke University Medical Center, Durham NC, USA
[3] Modus Medical Devices Inc. (IBA QUASAR), London, Ontario, Canada

*E-mail: kawtar.lakrad@duke.edu

Abstract. A comprehensive 3D dosimetry system consisting of the ClearView™ 3D dosimeter, VistaScan™ optical CT scanner, and VistaAce™ analysis software is commercially available and has the potential to be a strong verification tool with widespread applications. The clinical utility of this system (termed ClearView/Vista) has yet to be determined, and this work presents our first investigation. Clinical utility was evaluated using real patients' treatment plans: spine, head and neck, Single Isocenter Multi-Targets SRS (SIMT-SRS), and prostate cases. Independent validation was performed with a Monte Carlo-based dose calculation algorithm that was previously commissioned for clinical independent dose calculation. The analysis included line profile comparison (1D), isodose line comparison (2D), and global gamma analysis (3D). Results indicated a remarkable agreement between ClearView/Vista, the treatment planning system, and Monte Carlo, where gamma pass rates for all cases exceeded 94% even while using 2%2mm and a dose threshold of 10%. The ClearView/Vista 3D dosimetry system showed clinically acceptable accuracy and robust measurement and analysis in clinical settings.

1. Introduction

Advanced radiation therapy treatment modalities such as intensity-modulated radiation therapy (IMRT), stereotactic radiosurgery (SRS), adaptive therapy, and proton therapy have improved tumor treatment management and protected healthy tissues more effectively. However, radiation dosimetry accuracy must be verified more rigorously due to the increased complexity. 3D dosimetry is one possible technology that, given its comprehensive nature, is especially helpful, particularly for complex treatment plan validation (Oldham et al., 2012). An integrated 3D dosimetry system comprised of a 3D dosimeter ClearView™, an optical CT scanner Vista™16, and an analysis software VistaAce™ has been developed (IBA QUASAR, Canada). In the present study, we investigate the clinical applicability of this technology by testing actual treatment plans and innovative treatment techniques. In addition, we validate this comprehensive 3D dosimetry system using an independent Monte Carlo-based algorithm.

2. Materials and Methods

2.1 ClearView™ dosimeter

ClearView™ is a non-diffusing, radiochromic hydrogel dosimeter based on Tetrazolium Salts (TS), it features a radiochromic indicator dye that changes color from clear to purple when exposed to radiation. ClearView™ dose response is linear within a range of 0 Gy to 80 Gy. In addition, the

Journal of Physics: Conference Series **2799** (2024) 012011 doi:10.1088/1742-6596/2799/1/012011

updated formulation of ClearView™ is characterized by: energy and dose rate independence as well as signal stability over time (Penev et al., 2023). In this study four dosimeters were used, of which 3 were large size (15 cm diameter) and one was medium size (10 cm diameter).

2.2 Treatment Plans

An x-ray CT of a dummy dosimeter was acquired, using a GE Discovery RT590 CT scanner, for treatment planning purposes, and a TrueBeam STX Linear accelerator (Varian Medical Systems, Palo Alto, CA) was used for treatment delivery, while treatment planning was performed using the Eclipse treatment planning system (v15.6), with the dose being calculated using the Anisotropic Analytical Algorithm (v15.6.03). Four real patient treatment plans were recalculated on the dosimeter CT scan: a spine case, a head and neck, a Single Isocenter Multi-Targets SRS case (SIMT-SRS), and a prostate case. The prescribed doses were 30 Gy, 40 Gy, 40 Gy, and 30 Gy respectively, delivered on 1 fraction.

2.3 Imaging and analysis

Vista™ 16 is a cone beam optical CT scanner used to image 3D dosimeters. Vista™ 16 covers a wide scanning area (up to 15 cm diameter x 10 cm long) while maintaining a high spatial resolution (0.25, 0.5, 1.0, and 2 mm) (Olding et al., 2010). For data acquisition, the dosimeter is placed inside an aquarium filled with a refractive index matching fluid (degassed water), and a rotating stage is used to fix and rotate the dosimeter during data collection. The pre scans were acquired immediately before irradiation whereas the post scans were acquired 1 hour post-irradiation to allow signal stability. Image reconstruction was performed using an iterative reconstruction algorithm with a 1 mm resolution. Subsequently, the VFF (Visualization File Format) images were imported into the VistaAce™ software to perform manual image registration since the linear accelerator and the optical scanner have two separate coordinate systems. Optical attenuation to dose conversion was performed using a best-fit calibration between the planned and measured dose values. After calibration data was ready for dosimetric analysis, including line profile comparisons (1D), isodose line comparisons (2D), and global gamma analysis (3D).

2.4 Monte Carlo independent algorithm

SciMoCa (Radialogical LLC) is a dose calculation software that uses the Monte Carlo algorithm to perform secondary dose checks. Recently, it underwent comprehensive commissioning for complex radiotherapy plans including single isocenter multi-target intracranial radiosurgery (Adamson et al., 2023). The present study used SciMoCa (v1.7.1.5050) as a gold standard for comparing the Eclipse treatment planning system and the ClearView™ 3D dosimeter outputs.

3. Results and Discussion

Upon image reconstruction, it was apparent that the 3D dose distribution was similar to what was predicted by the treatment planning system as shown in Figure 1 (a, b, and c). A line profile comparison was performed including data from ClearView™, Eclipse, and Monte Carlo calculation algorithm for the spine case showing an agreement, as presented in Figure 1 (d), even while covering an area with different dose gradients (high=target volume, low=organ at risk "spine"). The isodose lines were well-aligned between the optical scanner and the treatment planning system as shown in Figure 1 (e), particularly in the central region where the isodose exceeds 60% of the prescribed dose. A dose-volume histogram (DVH) was generated to evaluate the dose constraints for both target volumes (Planning Target Volume and Clinical Target Volume) and the spinal cord as an organ at risk as shown in Figure 2.

Journal of Physics: Conference Series **2799** (2024) 012011 doi:10.1088/1742-6596/2799/1/012011

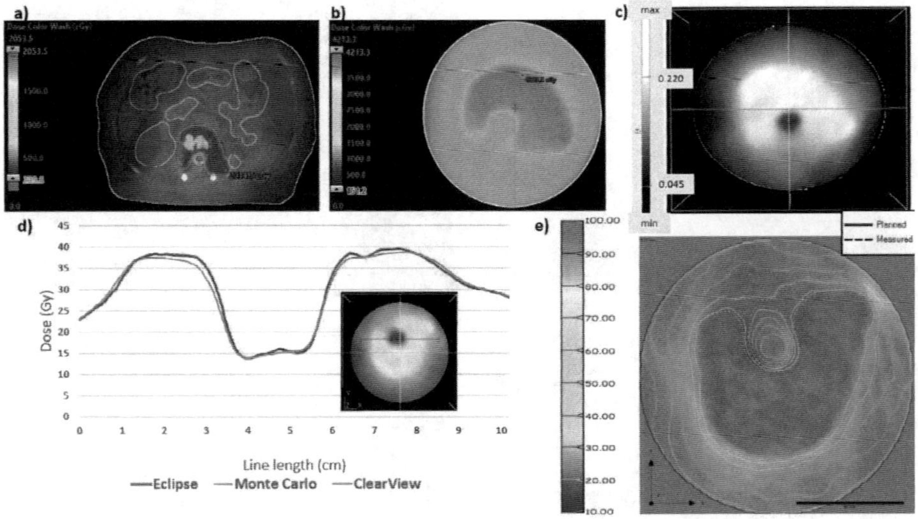

Figure 1. a) original spine treatment plan, b) spine treatment plan recalculated on the dosimeter geometry, c) image reconstruction after treatment plan delivery to the dosimeter, d) line profiles comparison, e) central axial slice isodose lines comparison (Eclipse=Planned vs ClearView=Measured).

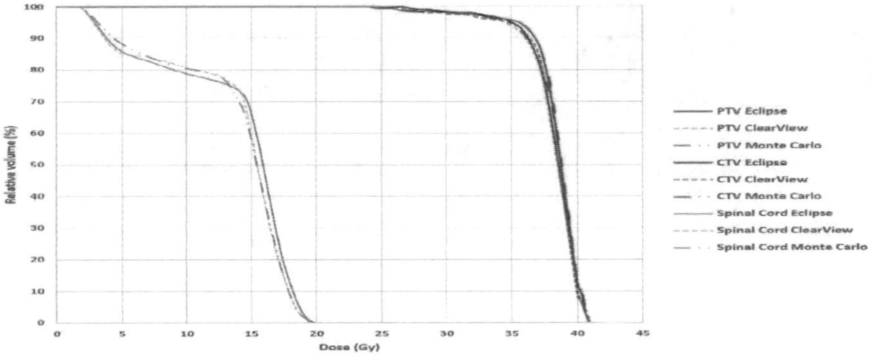

Figure 2. Spine Case dose volume histogram.

The DVH statistics for Eclipse, ClearView™, and Monte Carlo are slightly different, which can be attributed to the anatomical difference between the patient and the dosimeter container.

As demonstrated in Figure 3, the global gamma analysis indicated agreement for each clinical case even when strict gamma criteria such as 3%1mm, 5%1mm, and 2%2mm were used. The Gamma Pass Rates for the head and neck case were 97.5% (ClearView™ vs. Eclipse), 96.2% (ClearView™ vs. Monte Carlo), and 100% (Eclipse vs. Monte Carlo) using a 5%1mm criterion and a 10% dose threshold. Figure 4 displays 3D gamma maps that highlight the pass/fail region for the spine case. If a treatment plan is failing, these 3D gamma maps provide insight into where the failure is located. For instance, in this case, gamma fails within the high gradient region, specifically between 50% and 80% isodoses. It is possible that there was a slight misalignment in the registration along the edge of the Y-axis.

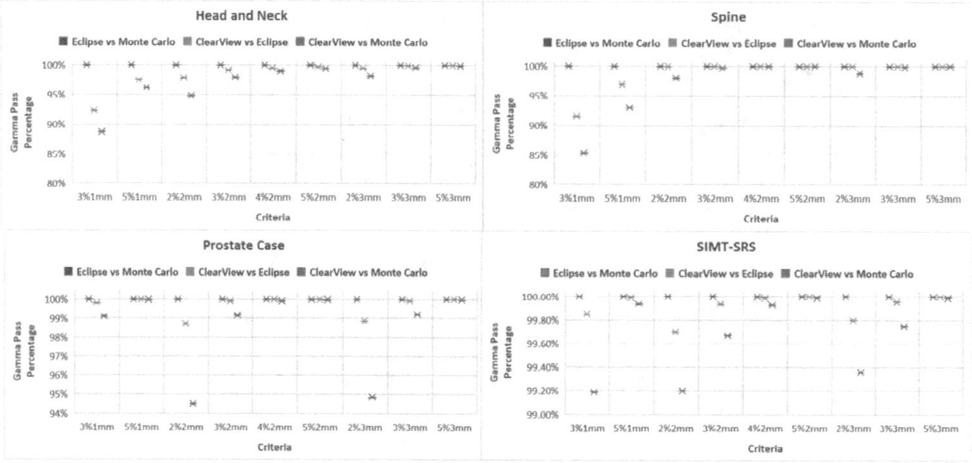

Figure 3. Scatter plots of the global gamma analysis for all irradiations

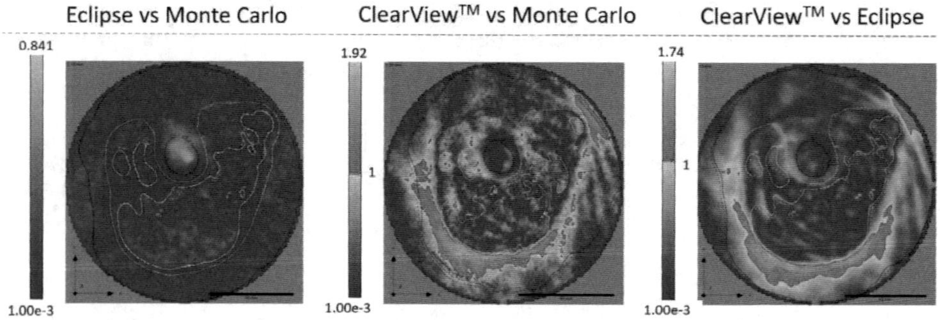

Figure 4. Gamma maps highlighting the pass (blue) and fail (red) regions, using a 3%1mm criterion and a 10% dose threshold

4. Conclusion

For a 2%2mm gamma criterion and a 10% dose threshold, the global gamma pass rates, for head and neck, spine, prostate, and SIMT-SRS cases, were 97.83%, 99.95%, 98.70%, and 99.70% respectively. This work demonstrated that the ClearView/Vista 3D dosimetry system has the potential for widespread adoption in the dosimetric verification of complex clinical and advanced radiation therapy treatment plans.

References

[1] Adamson, J., Erickson, B.G., Wang, C., Cui, Y., Alber, M., Kirkpatrick, J., Yin, F.-F., 2023. Generation, validation, and benchmarking of a commercial independent Monte Carlo calculation beam model for multi-target SRS. Zeitschrift für Medizinische Physik S0939388923000934.

[2] Oldham, M., Thomas, A., O'Daniel, J., Juang, T., Ibbott, G., Adamovics, J., Kirkpatrick, J.P., 2012. A quality assurance method that utilizes 3D dosimetry and facilitates clinical interpretation. Int J Radiat Oncol Biol Phys 84, 540–546.

[3] Olding, T., Holmes, O., Schreiner, L., 2010. Cone beam optical computed tomography for gel dosimetry I: Scanner characterization. Physics in medicine and biology 55, 2819–40.

[4] Penev, K.I., Mulligan, M., Mequanint, K., 2023. Optimization of the Dose Rate Effect in Tetrazolium Gellan Gel Dosimeters. Gels 9, 334.

Journal of Physics: Conference Series **2799** (2024) 012012 doi:10.1088/1742-6596/2799/1/012012

A heterogeneous-cavity model for miniaturized detectors of orthovoltage X-rays

F. Thevenet[1], S. Keshmiri[2], M. Moussaoui[1], B. Huffschmidt[2], J. Esteves[1], G-N. Lu[1], J-F Adam[2,3] and P. Pittet [1*]

[1] Universite Claude Bernard Lyon 1, INL, UMR5270, CNRS, INSA Lyon, Ecole Centrale de Lyon, CPE Lyon, 69622 Villeurbanne, France
[2] STROBE - Synchrotron Radiation for Biomedical Research UGA Inserm UA7, Saint Martin d'Hères, 38400, France
[3] Centre Hospitalier Universitaire Grenoble Alpes CHU-GA, Grenoble, 38043, France

*E-mail: patrick.pittet@univ-lyon1.fr

Abstract. To meet the needs of modelling small-sized detectors of orthovoltage X-rays, for spatially fractionated radiotherapy, we propose a model to determine responses of the detectors in cases of highly heterogeneous cavity. The model was applied to a detector whose cavity features a sparse high-Z perovskite scintillation layer embedded in an epoxy resin. Irradiations were carried out using the *Small Animal Radiation Research Platform*. Local photon and electron spectra were computed by the SpekPy toolbox and Penelope Monte Carlo simulations, respectively. The detector model considers that X-ray photons mainly interact with medium surrounding the scintillation layer, and that the output of the detector mainly results from secondary electrons in this layer. Results from model computations are compared with measurement data, showing differences less than 16% for beam energies in the 60 kVp -220 kVp range.

1. Introduction

Sub-millimeter fractionated high-dose-rate radiotherapy, such as MRT (Microbeam Radiation Therapy), uses orthovoltage X-rays and requires miniaturized detectors for dosimetry and Quality Assurance (QA). The detectors could be of different types, such as semiconductor devices operated on radio-electric effects, or scintillation structures based on radio-luminescence emission. Considering packaging of sensing elements and signal transmission schemes, a typical detector cavity – the part of the detector receiving X-ray radiation - is heterogeneous in terms of integrated materials of different components. In general, a detector cavity could integrate a signal sensing element in some cases, and a group of sensing elements in others. Predictably, the response of the detector, used to measure the average dose in a water voxel of the detector cavity volume, depends on its internal structure and material composition. As more and more new miniaturized detectors come to be employed, there has been demand for developing a model to quantify responses of heterogeneous-cavity detectors [1][2]. This is because small and large cavity theories are not applicable, and the intermediate cavity theory relies on similar media within the detector cavity, which is not the case for heterogeneous-cavity detectors. We propose here a new model to compute such responses, with the following assumptions: i) each sensing

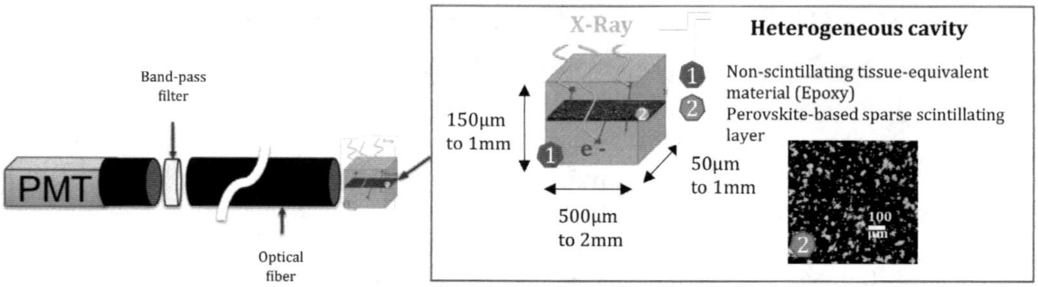

Figure 1. Schematic view of the detector with inset showing the heterogeneous cavity with a fluorescence image of micron-thick sparse scintillating layer

element in the cavity is sufficiently small that its direct interaction with the photon beam can be neglected and; ii) these elements are spaced apart (in terms of electron range) to prevent interaction between them.

As an example of model application and validation, we have studied a detector made of a micron-thick Perovskite-PDMS scintillating layer embedded in a tissue-equivalent organic material.

2. Materials and methods

2.1 Materials

The heterogeneous-cavity detector under this study consists of a micron-thick sparse CH5N2PbBr3 Perovskite-PDMS scintillating layer embedded in a tissue-equivalent organic material (epoxy) as illustrated in figure 1. This part is optically coupled to a 1 mm plastic multimode fiber. The scintillating signal at the fiber output is filtered by a band-pass filter (543 nm/22 nm) and then detected by a photon counting PMT Module. It is noted that Cherenkov emission can be neglected since the energy threshold in the PMMA and epoxy is higher than most kV energy spectrum values [2]. Measurements on the response of the detector under a 5 mm-diameter circular radiation field were performed using the *Small Animal Radiation Research Platform* (SARRP, Xstrahl, Camberley, UK) installed at IRMaGe (SAXO platform), in Grenoble, France. The detector cavity was positioned at a reference depth of 2 cm in a RW3 phantom with 4 cm of backscattering material. Reference dose measurements were carried out at the reference depth with a calibrated PinPoint 3D ionization chamber (PTW GmbH, Germany).

2.2 Methods

With consideration of specific experimental conditions, the X-ray spectra and fluencies from the SARRP tube were modelled by using SpekPy v2.0 software toolkit [3]. Determination of the local photon and electron spectra at the reference depth were based on the X-ray normalized energy spectra. Local photon spectrum for each beam energy (depending on the kVp tube setting) noted $\varphi_{2cm,kVp}(E)$ was determined by adding 2 cm of water filtration in SpekPy. The local photon fluencies were weighted by backscatter factors according to AAPM TG-61 protocol [4]. The local electron spectra noted $\Phi_{2\,cm,kVp}(E)$ were generated using the MC code Penelope [5]. For modelling the detector, we assume that in the heterogeneous detector cavity, X-ray photons interact primarily with the epoxy material under Charge Particle Equilibrium (CPE) conditions.

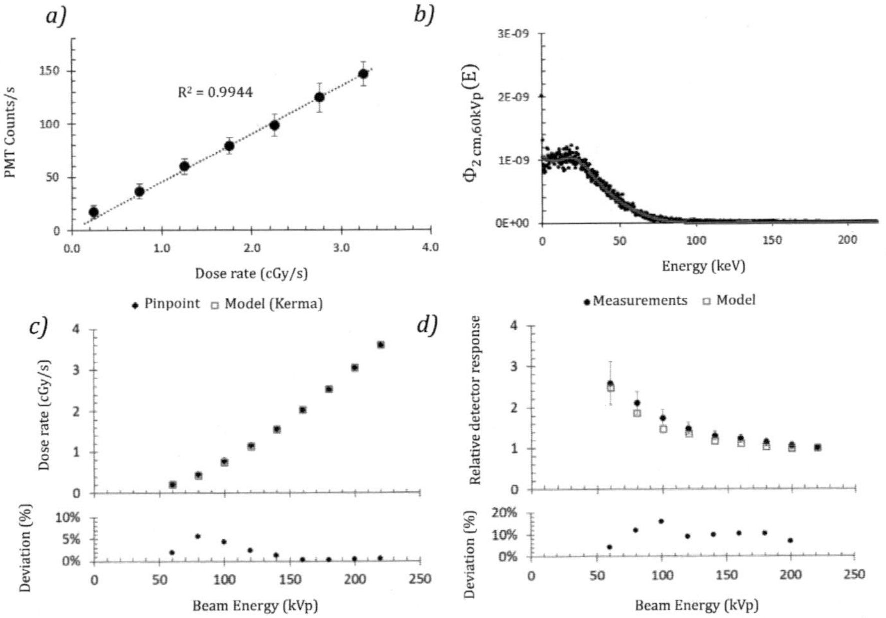

Figure 2. a) Linearity of the scintillation intensity as a function of dose at 2 cm in water for a beam energy of 220 kVp and with an additional filtration of 0.15mm of copper. Error bars correspond to ±1σ; b) Local electron spectra at reference depth for the 220 kVp beam energy; c) Computed kerma and measured dose at reference depth and d) measured and modelled detector response

Thus, the dose can be determined by the collision kerma in this material, which is very close to the kerma in water, $K_{w,kVp}(2\ cm)$. On the other hand, we also assume that the scintillation signal is mainly due to the interaction of secondary electrons with the sparse scintillation layer. Assuming that the local electron spectra is not modified by the presence of this thin layer, the scintillation signal is proportional to the restricted cema in Perovskite, $C_{Perov,kVp}(2\ cm)$ [6]. Based on these assumptions, the response of the detector can be determined as follows:

$$R_{kVp} = \frac{C_{Perov,kVp}(2\ cm)}{K_{w,kVp}(2\ cm)} = \frac{\left[\int_{\Delta}^{E_{max}} \Phi_{2cm,kVp}(E).\left(\frac{S_{\Delta}(E)}{\rho}\right)_{Perov} dE + TE\right]}{\int_{0}^{E_{max}} \varphi_{2cm,kVp}(E).\left(\frac{\mu_{en}(E)}{\rho}\right)_{w} .E\ dE} \quad (1)$$

where Δ is the energy threshold below which electrons are assumed to be absorbed locally in the scintillating layer, $\frac{S_{\Delta}(E)}{\rho}$ and $\frac{\mu_{en}(E)}{\rho}$ are respectively the restricted collision stopping power and mass energy-absorption coefficients based on the data from NIST, and TE is the track-end term related to the effect of finite delta ray ranges [6].

3. Results

Figure 2.a plots measured results on the response of the detector (PMT counts per seconds) irradiated at different dose rates ranging from 0.24 cGy/s to 3.25 cGy/s. A linear trend is observed confirming a linear response with no dose rate dependence. Figure 2.b shows computed the local electron spectrum for 220 kVp beam energy, indicating more significant contributions at lower energies, which can be attributed to slowing down of Compton electrons [7]. Local photon spectra and fluencies were used to compute the kerma in water at reference depth. Figure 2.c compares the obtained kerma with measured dose using a PinPoint ionization chamber. There is a good agreement, with differences within 6% over the range of 60 to 220 kVp, which confirms that the spectra and fluencies from the SARRP tube have been correctly modelled. Measured and modelled results on the detector response (counts/cGy) as a function of beam energy are plotted in Figure 2.d, with data points normalized to the 220 kVp beam energy. Deviation of the computed response from the measured one remains lower than 10% for [120 kVp, 220 kVp] energy range. It becomes larger (up to 16%) for [80 kVp, 100 kVp] range, which may be due to a lack of precision in spectrum modelling in this range, and/or increased measurement uncertainty at a low dose rate.

4. Discussion

The coherence between model computations and measurements is a confirmation for the model assumptions: X-ray photons interact mainly with the medium surrounding the thin scintillating layer, and the scintillation signal comes from irradiation of this layer by secondary electrons. These assumptions are applicable to other types of heterogeneous cavity detectors, such as those studied by Delage *et al.* [2]. It means that, by choosing a tissue-equivalent surrounding material, constraints on the tissue equivalence of the scintillating material can be relaxed, since the energy dependence of the stopping powers is much weaker than that of the mass energy-attenuation coefficients. This proposed model can provide an alternative approach to intermediate cavity theory without the need to choose d-parameter (a weighting factor depending on the cavity size: from unity to zero with the increase of cavity dimensions) that does not meet with consensus [1]. The presented study was focused on the response of the detector in terms of energy dependence and measurements were performed at low dose rates, whereas this detector was designed for FLASH therapies. Preliminary tests carried out under MRT conditions at ESRF (not shown here) confirmed a linear response of this detector at very high dose rates (a few hundred to a few thousand Gy/s), and thus its suitability for this application.

5. Conclusion

We have developed a heterogeneous-cavity model for small-sized detectors. It was applied to a detector based on a thin scintillating layer of Perovskite embedded in epoxy resin. The model allows computations of the response of the detector. By comparing model computations with measurements for beam energies in the 60 kVp -220 kVp range, differences remain lower than 16%, and could be narrowed by further studies such as more accurate local spectral modelling.

Acknowledgements

With financial support from ITMOCancer AVIESAN (Ioda-Master project) and from the ANR (ANR-11-LABX-0063, ANR-19-CE19-0021). Experimental developments were carried out at Nanolyon (Renatech+ network) and IRMaGe platforms.

References

[1] A. Vaniqui, B. R. Walters, G. P. Fonseca, and F. Verhaegen, 2019 *Phys. Med. Biol.* **64** 165001
[2] M. E. Delage, M. E. Lecavalier, D. Lariviere, C. N. Allen, and L. Beaulieu 2019 *Phys. Med. Biol.* **64** 095027
[3] G. Poludniowski, A. Omar, R. Bujila, and P. Andreo 2021 *Med. Phys.* **48** 3630–3637
[4] C. M. Ma *et al.* 2001 *Med. Phys.* **28** 868–893
[5] F. Salvat 2019 OECD Publishing Paris NEA/MBDAV/R(2019)1 https://doi.org/10.1787/32da5043-en
[6] G. H. Hartmann, P. Andreo, R. P. Kapsch, and K. Zink 2021 *Med. Phys.* **48** 7461–7475
[7] Å. C. Tedgren and G. A. Carlsson 2013 *Phys. Med. Biol.* **58** 2561–2579

Journal of Physics: Conference Series **2799** (2024) 012013 doi:10.1088/1742-6596/2799/1/012013

Scintillation Imaging as Versatile Tool for Validation of UHDR Treatment Plans Across Modalities

Megan Clark[1], Roman Vasytlsiv[1], David Gladstone[1,2], Anthony Mascia[3,4], Joseph Harms[5], Petr Bruza[1]

[1] Thayer School of Engineering Dartmouth College, Hanover NH, USA
[2] Dartmouth Cancer Center, Lebanon, NH, USA
[3] Cancer and Blood Diseases Institute, Cincinnati Children's Hospital, Cincinnati, Ohi, USA
[4] Department of Radiation Oncology, College of Medicine, University of Cincinnati, Cincinnati, OH
[5] University of Alabama at Birmingham, Department of Radiation Oncology, Birmingham, AL

*E-mail: megan.a.clark.th@dartmouth.edu

Abstract. Scintillation imaging has been shown to provide high spatio-temporal surface dose information for delivery verification of ultra-high dose rate (UHDR) beams. To obtain higher dose rates, modified clinical machines are often utilized, requiring extensive re-characterization and treatment validation. The goal of this work was to demonstrate the utility of 1.) projecting surface images to depth and 2.) quinine-doped water tank imaging for 3D beam verification across different UHDR treatment modalities.

1. Purpose

The rapid push towards clinical translation of ultra-high dose rate radiation therapy (UHDR-RT) necessitates dosimetry tools for characterization and validation of beams delivered at dose rates exceeding 40 Gy/s. In attempt to achieve the coveted FLASH effect, treatment plans will not only be optimized using the traditional dose volume histogram but also the dose rate volume histogram. However, existing technologies lack the ability to concurrently capture spatially and temporally resolved beam delivery at the resolution necessary to accurately reconstruct the complex UHDR dose and dose rate maps over the entire treatment volume. Additionally, investigation into relevant treatment parameters to elicit the FLASH effect require dosimeters that span delivery conditions and treatment modalities.

The FLASH Radiotherapy for the Treatment of Symptomatic Bone Metastases (FAST-01) clinical trial was the first translation of UHDR-RT to a clinic in the United States[1], using a clinical proton pencil beam scanning (PBS) cyclotron converted to UHDR mode via increased beam current and shoot-through transmission. Many studies have used UHDR PBS fields for investigation of the FLASH effect, with more recent, complex treatment fields being produced with updated treatment planning systems. UHDR electron beams capable of eliciting the FLASH effect are also under investigation. For both modalities, the machine deliveries, log files and treatment planning systems will require validation for safe clinical implementation, and full documentation of treatment deliveries, including beam structure, will be required for clinical outcome analysis.

Previous work has demonstrated the utility of fast scintillation imaging for validation of dose and dose rate at the surface and for depth dose profiling of UHDR beams[2,3]. The goal of this work was to use the characterized imaging tool to 1.) provide validation of the machine log files used to analyse the dose rate delivered in the FAST-01 clinical trial fields and complex UHDR PBS treatments by projecting surface scintillation images to depth and 2.) to characterize UHDR-capable accelerators via Abel-reconstructions of scintillation water tank imaging.

2. Methods

2.1 Validation of FAST-01 Clinical Treatment Fields and Complex UHDR PBS Plans via Imaging Depth Projection

An ultra-fast intensified CMOS (iCMOS) camera was set up to image a scintillator target (Rapidex Medium, Scintacor, Cambridge UK) to which UHDR beams were delivered by a Varian ProBeam cyclotron system, as described and validated previously and shown in Figure 1a[2]. The camera system acquired images of scintillation response to PBS proton fields at a 4.5 kHz sampling rate and the light intensity was calibrated to dose with film. The system log files reported on total irradiation time and dose and dose rate to isocenter, which was located at 5 cm depth for this setup configuration, requiring that the images of surface dose be projected to 5 cm depth for direct comparison. To project the surface images to depth, each 2D image in the stack was convolved with a measured profile spread of a single spot delivered at 250 MeV, 210 nA at 5 cm depth and the cumulative intensity was scaled by the ratio of the measured percent depth dose curve at 5 cm to that at 0 cm for each frame in the image stack. Clinical treatment plans for the predefined FAST-01 fields (7x7, 7x10, 7x12, 7x14, 7x16, 7x18,7x20 cm^2) and a clinically realistic lung treatment plan were all delivered at over 65 Gy/s PBS dose rate[1,4]. To compare dose rates measured with the imaging system and diode to that from the log files, the threshold dose used in PBS dose rate, d_t, was chosen at 0.27 Gy, for the data both at surface and 5 cm depth. This threshold choice captures the inherent noise in the imaging and diode recordings while still accurately reflecting the dose rates measured by the log files, such that 0.01 Gy in the log file corresponded to 0.27 Gy. The impact of the chosen threshold on dose rate was investigated using these two thresholds as comparisons.

2.2 Beam Characterization using Quinine-Doped Water Tank Imaging

Lateral projected 2D fluorescence distributions of UHDR deliveries were imaged in a quinine-doped water tank, using a fast, intensified CMOS camera, as shown in Figure 2.a, fitted with a 450 nm bandpass filter to remove the energy-dependent Cherenkov signal. Images of an electron (9 MeV, 6cm field, Mobetron IntraOP) and the cyclotron proton beam (described in Section 2.1) were collected, and an Abel transform was used to provide a 3D reconstruction from the 2D distribution, assuming an axial symmetric dose distribution from previous film measurements, allowing analysis of on-axis depth-dose profiles for the cumulative beams[3]. For characterization of the Mobetron, imaged profiles, namely D_{max}, and R_{50}, were compared between conventional and UHDR delivery and percent-depth-dose measurements taken with an EDGE-diode detector. Depth dose profiles of the UHDR cyclotron single spot were compared at two energies to demonstrate feasibility.

3. Results

3.1 Validation of FAST-01 Clinical Treatment Fields and Complex UHDR PBS Plans at Depth

Camera response was shown to be linear with dose (R^2 =0.9998) and current (R^2=0.9883) from 2-12 Gy and 20-210 nA, respectively. Dose and PBS dose rate were calculated with the imaging data at the surface and at 5 cm depth for each FAST-01 field delivered. While

PBS dose rate was evaluated for the entire area of the field, as shown in Figure 1b, the dose rate at the center of the field was compared to the diode and log files recordings. At the surface, the average imaged PBS dose rate across the seven fields was 114±1 Gy/s and agreed with the diode recordings within 1±1%. The dose at the center of the field at 5cm depth from the projected images agreed with the reported dose (from log files) within 0.13±0.03 Gy (2±1%), with a mean imaged dose of 8.4 Gy. The average PBS dose rate from projected images at 5 cm depth was 62±1 Gy/s which agreed with the reported values within 2±3%. For the simulated lung field, dose was 5% higher in the depth imaging data compared to the surface imaging, and the PBS dose rate map varied with threshold dose rate chosen, as shown in Figure 1c.

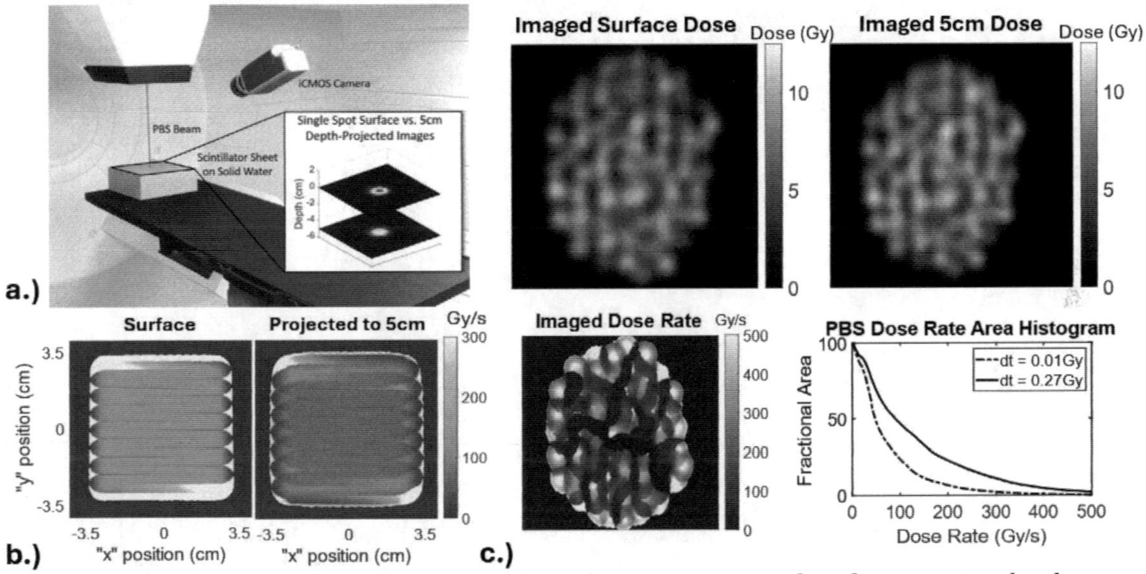

Figure 1. a.) Camera setup at UHDR-capable cyclotron centers and surface imaging depth projection from flat scintillator sheet surface imaging on solid water. **b.)** PBS dose rate maps at the surface compared to 5 cm depth calculated with projected surface imaging for FAST-01 7x7 cm^2 field. **c.)** PBS surface dose rate map for clinically realistic lung field delivered at UHDR dose rates and dependence of PBS dose rate area histogram on PBS threshold dose.

3.2 Beam Characterization using Quinine-Doped Water Tank Imaging

Imaging depth dose profiles of the 9 MeV Mobetron beam under conventional and UHDR (pulse repetition frequency of 120 Hz across various pulse widths) conditions were compared to an EDGE diode (SunNuclear), as shown in Figure 2c. D_{max} was measured to be 20.1 mm and 20.2 mm with the diode and imaging system, respectively, for delivery in conventional mode. Under UHDR conditions, the imaging system measured D_{max} of 19.1±0.2 mm, whereas the diode reported a D_{max} of 20.8±1.2 mm. R50 agreed between the imaging and diode with 37.5±1.2 mm and 37.7±1.1 mm, respectively. These profiles also confirmed a beam-specific dependence on pulse width, likely due to change in beam energy and scatter, an important consideration for treatment planning. Depth profiles of a PBS single spot delivered at 97 MeV and 103 MeV confirmed the ability to use the same methodology for proton range verification.

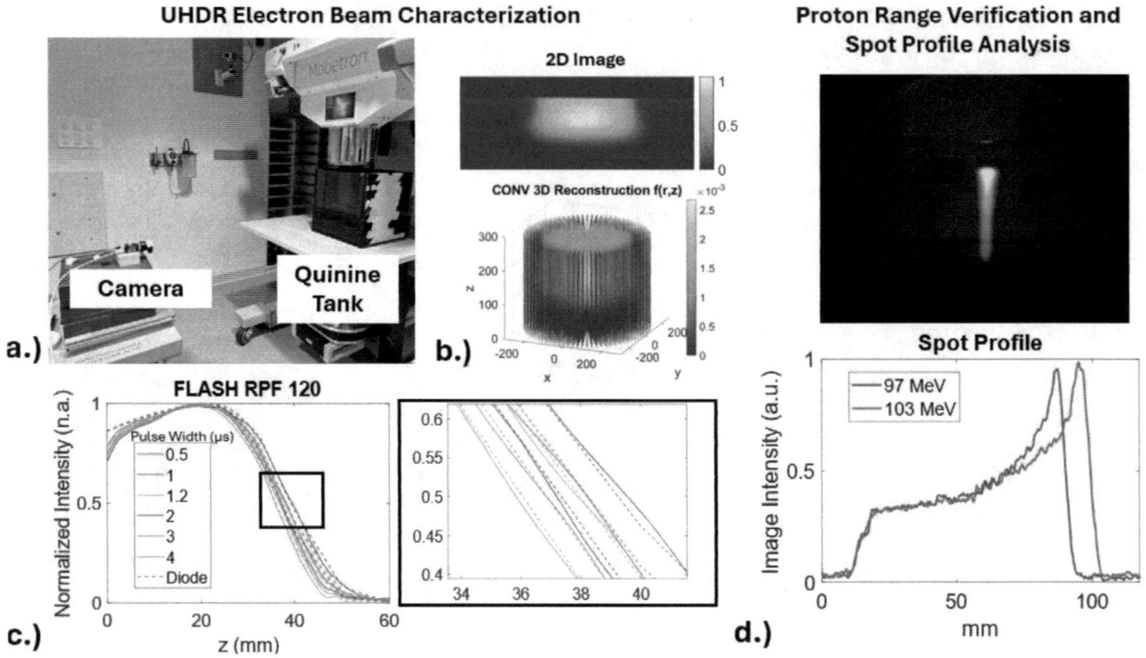

Figure 2. a.) Camera setup to image quinine-doped water tank under Mobetron UHDR-capable electron system. **b.)** Demonstration of Abel Transformed 2D image to create 3D reconstruction of delivered beam. **c.)** Percent depth-intensity curves comparing UHDR repetition pulse frequency (PRF) 120Hz for different pulse widths between the imaging and EDGE diode. **d.)** Single spot UHDR image and profile compared between 97 MeV and 103 MeV.

4. Conclusion and Next Steps

This study demonstrates the ability to use surface imaging, either projected to depth or Abel transformed, for 3D dose and dose rate verification under different deliver conditions. The versatility of remote scintillation imaging offers solutions for comprehensive UHDR pre-treatment QA, UHDR beamline systems validation, as well as in vivo patient monitoring and dosimetry.

Acknowledgements

NIH R44 CA268466, NIH U01 CA260446, and this work used the radiation shared resource of 5P30CA023108-40 and DoseOptics Cameras

References

[1]Mascia AE, Daugherty EC, Zhang Y, et al. Proton FLASH Radiotherapy for the Treatment of Symptomatic Bone Metastases: The FAST-01 Nonrandomized Trial. *JAMA Oncol.* 2023;9(1):62. doi:10.1001/jamaoncol.2022.5843

[2] Clark M, Harms J, Vasyltsiv R, et al. Quantitative real-time measurements of dose and dose rate in UHDR proton pencil beams via scintillation imaging system. Published online 2023. doi:10.48550/ARXIV.2312.14841

[3] Ashraf MR, Rahman M, Zhang R, Cao X, Williams BB, Hoopes PJ, Gladstone DJ, Pogue BW, Bruza P. Technical Note: Single-pulse beam characterization for FLASH-RT using optical imaging in a water tank. Med Phys. 2021 May;48(5):2673-2681. doi: 10.1002/mp.14843. Epub 2021 Mar 31. PMID: 33730367; PMCID: PMC10771323.

[4] Folkerts MM, Abel E, Busold S, Perez JR, Krishnamurthi V, Ling CC. A framework for defining FLASH dose rate for pencil beam scanning. Med Phys. 2020;47(12):6396-6404. doi:10.1002/mp.14456

Journal of Physics: Conference Series **2799** (2024) 012014 doi:10.1088/1742-6596/2799/1/012014

Polymer gel dosimetry of added High-Z elements in radiation fields of electrons and photons with low and high photon energy

Katharina Buchner[1], Manuela Gober[2], and Andreas Berg[3,4]

[1] Department for Radiation Oncology, Hospital Favoriten, Vienna, Austria
[2] Department of Radiation Oncology, Medical University of Vienna, Vienna, Austria
[3] Center for Medical Physics and Biomedical Engin., Medical University of Vienna, Austria
[4] High-Field MR-Center; Medical University of Vienna, Vienna, Austria

*E-mail: andreas.berg@meduniwien.ac.at

Abstract. A dose enhancement in tumors can in principle be achieved using an enrichment of elements with high atomic number ("High-Z therapy"). The proof of dose enhancement at low non-toxic concentrations remains difficult with standard dosimeters due to the small distance of secondary electrons of X-ray photon energies. We investigated within an exploratory research approach the dose enhancement in tissue equivalent environment using MRI analysis of polymer gel dosimeters doped with molecules containing high-Z elements: bismuth, silver iodide and iopromide, a radio-diagnostic contrast agent for X-ray imaging, in photon fields of low (200 kV X-ray) and high energy (6 MV) and in an electron field (6 MeV). An enhancement of the dose response up to a relative dose enhancement factor $rDE_{exp} \cong 1.4$ was observed for low concentrations (w/w=0.006) of iopromide, already approved for vascular angiography with X-rays. Experimental limitations and challenging interpretations are outlined.

1. Introduction

The dose sensitizing effect (DSE) of added metallic (e.g., gold, rare earth nanoparticles) and non-metallic elements with high atomic number (Z) at concentrations below 1% has been proved by Monte-Carlo simulations, in-vitro and pre-clinical investigations [1-5]. DSE has mainly been shown for low energy (kV) X-rays, but has also been reported for high (MV) X-rays as used in clinical External Beam Radiation Therapy (EBRT) in vitro on cell systems for bismuth oxide and tantalum pentoxide [1]. These findings represented the basis for the proposal to enrich tumor tissue by nano-particles containing elements with high Z before irradiation ("high-Z therapy"). Polymer and Fricke gel dosimetry [6] appears to be well suited for the measurement of a physical dose enhancement in close vicinity of such high-Z nanoparticles due to their inherent tissue equivalence and potential for high-resolution 3D-dosimetry. Relative dose enhancement has been proved already for n-PAG and Fricke gels containing iodine and barium [3-5] using Synchrotron irradiation at low photon energy (33-80 keV). Also, dose enhancement close to thin metal layers made of lead has been demonstrated as a result of the large range of secondary electrons produced in an 18 MV photon beam [7]. Physico-mathematical simulations on the relative dose enhancement (rDE_{th}), based on the mass energy absorption coefficient [4,8] do indicate, that

Journal of Physics: Conference Series **2799** (2024) 012014 doi:10.1088/1742-6596/2799/1/012014

significant rDE_{th} at low concentrations cannot be expected for high photon energies as used in clinical EBRT. We investigated within an exploratory research approach the dose response in MAGADIT polymer gels [9] using Magnetic Resonance (MR-) micro-imaging for the high-Z dopants: bismuth (Z_{Bi}=83), silver iodide, known for its sensitivity to irradiation from dosimetric films and iopromide (Z_I=53), a radio-diagnostic contrast agent used for X-ray imaging of the vessel distribution. The relative dose enhancement (rDE_{exp}) was measured with MRI as ratio of the dose sensitivities $a_i=\Delta R2/\Delta D_i$ with and without dopants to the polymer gel. It is reported in photon fields of low (200 kV), but also high (6 MV) energy and in an electron field (6 MeV) with no expected rDE_{th} in order to prove the validity of the theoretical predictions.

2. Materials and Methods

2.1 Investigational approach and theoretical modelling

This explorative approach on dopants to polymer gels for dose enhancement is based on the optimization of the relative ratio of the energy absorption coefficients (μ_{en}/ρ) [8] at X-ray energy E of the element to be investigated and water, assumed to be relevant for most tissues (eq. 1)[4]:

$$rDE_{th} = \frac{\left(\mu_{en}/\rho\right)_E^{Z-doped\,gel}}{\left(\mu_{en}/\rho\right)_E^{gel}} = \frac{w_z\left(\mu_{en}/\rho\right)_E^{Z}+(1-w_z)\left(\mu_{en}/\rho\right)_E^{gel}}{\left(\mu_{en}/\rho\right)_E^{gel}} \qquad (1)$$

w_z denotes here the relative mass contribution of the element with atomic number Z to the overall mass of the doped polymer gel. The rDE_{th} could only be estimated, as the energy spectrum of the 200 kV tube was unknown. The average energy of the photon field (E_{max}=200kV) is approximated to be about 1/3 of maximum energy. Actually, for an estimate of the rDE_{th}, the energy absorption coefficient μ_{en}/ρ for photons with energy $<E> \cong 50$ keV was considered and used as available from the NIST database [8]. The estimated rDE_{th} according to equ. 1 and table 1 for the different relative mass contributions w_z for Ultravist (U1, U2) and Boron powder (B1) was determined to be: $rDE_{U1}\cong 2$; $rDE_{U2} \cong 3.8$; $rDE_{B1}\cong 1.5$. Different co-factors with potential impact on the MR detected dose response have to be considered: a) the potential chemical interactions with the monomer source material (methacrylic acid), b) the potential interference of the high-Z ingredients with the dose sensitive physico-chemical interactions during radical initiated polymerization and c) direct impact e.g., by magnetic susceptibility on the MR-detection principle relying on the dose response detected by the transverse relaxation rate R2=1/T2 [6,9]. Moreover, we strived for chemical compounds as less toxic as possible.

2.2 Materials, irradiation, Magnetic Resonance Imaging (MRI) and data processing

The procedure for the manufacturing of the MAGADIT polymer gels is described in [9]. This procedure was followed with one main change: we prepared the norm-oxic polymer gels with

Table 1. Ingredient concentrations for the doped MAGADIT polymer gels (Ultravist™300 with iopromide containing 300 mg iodine in 1 mL solution) and Bismuth powder indicating the relative mass concentration of the dopants with reference to the final weight of the doped polymer gel.

Ingredient	Gelatine	Distilled water	Methacrylic acid	Dithiothreitol	Iodine in iopromide U1/U2	Bismuth B1/B2
Relative mass (w/w)	16%	75%	8%	2 mmol/kg	0.6%/1,8%	0.3%/0.9%

gelatine at higher concentration (16%) due to the lower gel strength (240 Bloom) of the available porcine gelatine (table 1). The polymer gel was filled in small PET-containers (V≅32 mL) after adding and mixing the doping agents. It solidified during cooling to room temperature. For irradiation at low energy (200 kV) a preclinical radiation unit (Yxlon MAXISHOT Hamburg/GER) was used. The nominal dose levels in water were adjusted from precedent measurements on the dose rate using a vented semi flexible ionization chamber (PTW/Germany) in a water filled PET-container at similar irradiation conditions. The dose sensitivity was measured as slope ($a=\Delta R2/\Delta D$) of the relaxation rate $R2=1/T2$ increasing with nominal incident dose D. MRI was performed on a human high-field (B=7T) MR-scanner [9] (TR/TE=6620/20 ms, nr. echoes: 15, VS: 1.56x1.56x3 mm³). The rDE_{exp} was evaluated as ratio of the dose sensitivity (slope a) for the doped to the non-doped polymer gel ($rDE_{exp}=a_{doped\text{-}gel}/a_{non\text{-}doped\ gel}$). Errors for rDE_{exp} are estimated from the standard errors for the relevant slopes a in regression analysis (see fig. 1).

3. Results

The first two batches did indicate that the silver iodide (AgI) recipe resulted in a clumping of the tiny AgI flakes, which could also be observed in MR microimages as tiny dark spots. Though a dose response was observed, there had been no indication of a dose enhancement by AgI as indicated by an increased change of R2 with dose. The polymer gel with higher bismuth concentration (w_{B2}=0.9%) could not be manufactured due to increased rigidity, assumed to indicate early polymerization due to chemical interaction with bismuth without irradiation. Polymer gel recipe B1 did show a slight increase in dose sensitivity (rDE_{exp}≅1.2) compared to the non-doped polymer gel, though exhibiting a higher offset $R2_0$ at D=0 Gy indicating already some slight impact on the rigidity of the polymer gel network (fig. 1d). Adding iopromide as a clinically approved contrast

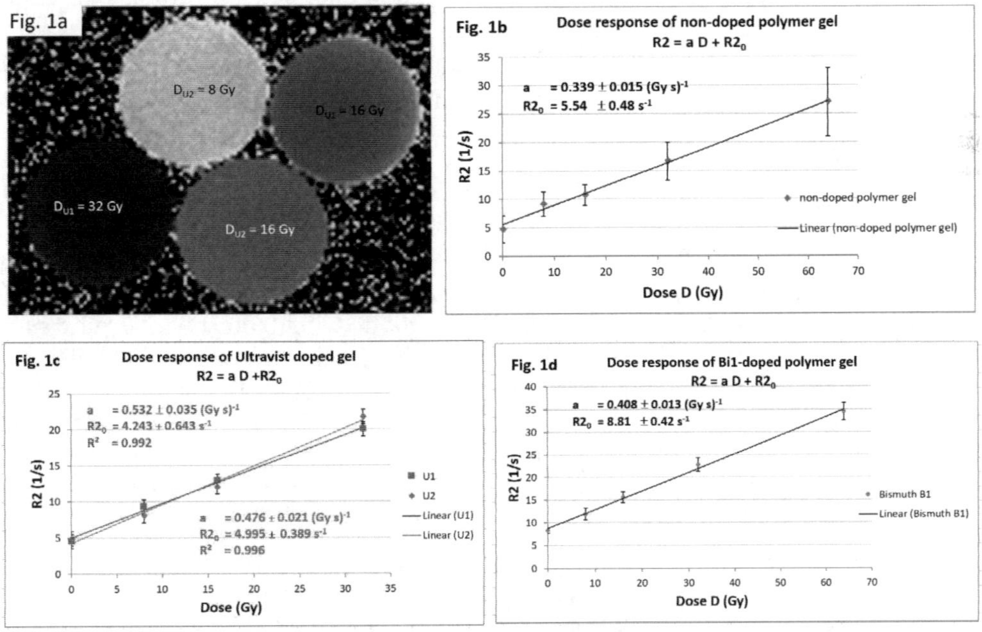

Figure 1. Dose response ($R2=T2^{-1}$) of MAGADIT polymer gels in an X-ray-field (200 kV): **a)** T2-maps of the polymer gels doped with iopromide **b)** dose response $R2=1/T2$ without added agents; **c)** with iopromide U1(w_U=0.6%) and U2 (w_{U2}=1.8%); **d)** bismuth (w_{Bi1}=0.3%). Note the increase in sensitivity (slope a) from undoped gel to doped gel (rDE_{U1}=1.40±0.09; rDE_{U2}=1.57±0.12 and rDE_{B1}=1.20±0.06.

agent (Ultravist™ solution) at low iodine concentration (w_{U1}=0.6%) resulted in a relative dose enhancement of rDE_{U1}≅1.4 (fig. 1c). Elevated concentration (w_{U2}=1.8%) only slightly increased dose sensitivity additionally to rDE_{U2}≅1.56.

4. Discussion

A dose enhancement is observed for the high-Z dopant iodine applied as contrast agent Ultravist™ at low concentrations for an X-ray radiation field (200 kV). The rDE_{exp} appears to be a little less than that expected from theoretical estimations (rDE_{th}≅2) based on the relative mass energy absorption coefficients (equ. 1) assuming a relevant photon energy of 50 keV for the X-ray tube. Also, a higher concentration of the High-Z dopant Ultravist™ did not significantly enhance rDE_{exp} additionally as predicted by the simple theoretical model resulting in equation (1). The reasons remain speculative without further systematic investigations, e.g., physico-chemical interactions between the generated radicals, iodine and the acrylic monomer resulting in non-linear correlation between the iodine concentration and detected polymerization. The results do not indicate a potential failure of the theoretical modeling of the rDE by high Z-dopants or restrict the potential of "High-Z therapy". Boudou reported for sodium iodide powder (w_1≅0.5%; w_2≅1%) as dopant for a Synchrotron source with 35 and 80 keV a dose enhancement between 1.51 and 1.97 [3]. Silver iodide did not show any dose enhancement effect, likely due to a clustering of the crystals to small particles and thus limiting the dose enhancement by low-range secondary electrons to the close environment of the accumulated crystals. Thus, it is also indicating the relevance of a homogeneous distribution of the high-Z elements in the targeted volumes for efficient dose enhancement. The early polymerization at higher concentrations of bismuth, even as low as 0.9%, demonstrated the potential impact of added agents on chemical detectors relying on the polymerization process and thus potential sources of dosimetric errors. No dose enhancement is observed for high-energy photons and electrons typical for EBRT. This observation confirms the theoretical modelling for the physical deposition of dose originating from the small differences in mass energy absorption coefficients at these high photon energies, as the Compton effect is dominant in this region and rather independent of the atomic number Z.

5. Conclusion

A relative dose enhancement in polymer gels could be observed for the dopants bismuth (Bi) and iodine (I) at concentrations below 1% for photon radiation fields at low energies as used for orthovoltage radiation therapy. There had been no indication of a physical dose enhancement at high photon energies (E=6 MV) and for electron irradiation (6 MeV), common in EBRT. Iodine applied as iopromide, available as contrast agent for X-ray clinical diagnostic imaging of vessels, offers potential for an increase of physical dose around tumours exhibiting an increased vessel support, e.g., due to an enhanced Vascular Endothelial Growth Factor (VEGF) of tumour cells.

References

[1] Choi J, Kim G, Cho SB *et al* 2020 *J Nanobiotechnol.* **18** 122 DOI: 101186/s12951-020-00684-5
[2] Matsudaira H, Ueno AM, Furuno I 1980 *Radiat. Res.* **84**(1):144–8 DOI: 10.2307/3575225
[3] Boudou C et al. 2007 *Phys. Med. Biol.* **52**(16):4881-92 DOI: 10.1088/0031-9155/52/16/011
[4] Corde S, Adam J-F and Biston MC 2005 *Rad. Prot. Dosim.* **117**(4): 425-31DOI: 10.1093/rpd/nci313
[5] Gastaldo J, *et al.* 2008 *Eur. J. Radiol.* 2008 **68**(3 Suppl.):S118-20 DOI: 10.1016/j.ejrad.2008.04.053
[6] Baldock C *et al* 2010 *Phys. Med. Biol.* **5** R1-R63 DOI: 10.1088/0031-9155/55/5/R01
[7] Warmington LL et. al. 2016 *Med. Phys.* ***43(12):6525*** DOI: 10.1118/1.4967483
[8] NIST National Institute of Standards, Physical Reference Data. *X-ray mass coefficients Table 3* [Online] https://physics.nist.gov/PhysRefData/XrayMassCoef/ElemTab/z53.html [ass: 18.03 2024]
[9] Khan M, et al. 2019 *Polymers* **11**(10): 1717 DOI: 10.3390/polym11101717.

Journal of Physics: Conference Series **2799** (2024) 012015 doi:10.1088/1742-6596/2799/1/012015

Applications of Polarization Imaging for Conventional and FLASH Radiotherapy Dosimetry

Émily Cloutier[1], Arthur Lalonde[2,3], Karim Zerouali[2], Luc Beaulieu[4,5] and Louis Archambault[4,5]

[1] Faculty of Science and Engineering, Université Laval, Québec, Canada
[2] Centre Hospitalier de l'Université de Montreal (CHUM), Montreal, Qc, Canada
[3] Université de Montréal, Montréal, Canada
[4]Department of physics, Université Laval, Québec, Canada
[5]Oncology division, CHU de Québec-Université Laval Research Center, Québec, Canada

E-mail: emily.cloutier@fsg.ulaval.ca

Abstract. The application of Cherenkov radiation in radiation therapy dosimetry has been limited by the anisotropic nature of the signal. Recently, polarization imaging was investigated as a method to correct Cherenkov anisotropy and allow precise dose measurements directly in a water tank. The aim of this study is to present polarization imaging as method for the measurement of ultra-high dose rate (UHDR) intra-operative electron beams. In this new approach, the polarized Cherenkov signal was isolated and utilized as a surrogate to evaluate the quality and consistency of both UHDR and conventional electron beams. Percent depth Cherenkov signal were measured for different energies, field sizes and dose rates. The results demonstrate high linearity ($R^2 > 0.99$) of the Cherenkov signal with the number of pulses and pulse width. The wide dynamic range of the device enabled measurement for both conventional and UHDR radiation beams making it a promising candidate for real-time quality assurance devices.

1 Introduction

Cherenkov radiation, first observed by Marie Curie, has historically been observed in nuclear detectors due to its distinctive blue glow emitted when charged particles travel through a medium at speeds greater than the speed of light in that medium [1]. However, its application in clinical radiation therapy for dose measurements has been limited due to the anisotropic nature of the Cherenkov signal, which was often treated as a spurious stem signal that needed to be removed [2, 3]. Early efforts to overcome this limitation included the addition of a fluorophore, such as quinine sulfate dihydrate [4, 5], and the calculation of correction factors using Monte Carlo simulations [6, 7]. Recently, the investigation of polarization imaging, have opened new avenues for utilizing Cherenkov radiation in radiation therapy dosimetry [8, 9]. Polarization, a fundamental property of light alongside its wavelength, carries valuable information that can be use to improve measurements precision.

Polarization imaging is a method that involves imaging a signal from different polarization axes to effectively isolate the polarized signal from the randomly polarized signal. Cherenkov radiation is inherently polarized, as it is produced from dipoles oscillating in a direction perpendicular to the trajectories of charged particles propagating in a dielectric medium. However, it was previously shown that Cherenkov radiation resulting from secondary electrons set in motion by radiotherapy treatment beams produces a signal that is partly polarized [9]. This is caused by the tortuous path of electrons that leads to Cherenkov signal being partly averaged into a randomly polarized signal when imaged by a camera.

Analysers, i.e polarizer that serves to measure the signal, are used to extract the polarization of the signal. According to Malus' law, the intensity of light transmitted through an analyzer is proportional to the square of the cosine of the angle between the transmission axis α_0 and the polarization angle of the incoming light, α:

$$I = I_0 \cos^2(\alpha_0 - \alpha) \tag{1}$$

When randomly polarized light, such as that emitted by scintillators, passes through the analyzer, its intensity decreases by half of its initial value. Malus' law can be modified to account for partially polarized signals, such as Cherenkov radiation:

$$I = I_{pol} \cos^2(\alpha_0 - \alpha) + I_{\text{Rand.pol.}} \cos^2(\alpha_0 - \alpha). \tag{2}$$

Here the mean value of $\cos^2(\alpha_0 - \alpha)$ for the randomly polarized portion of the signal is $1/2$. Consequently, neglecting transmission losses, equation 2 can be expressed as:

$$I = I_{pol} \cos^2(\alpha_0 - \alpha) + 0.5 \cdot I_{\text{Rand.pol.}}. \tag{3}$$

where I_0 corresponds to the intensity of the polarized beam. Hence, the non-polarized signal acts as a DC component that adds to equation 1. Solving equation 3, for each pixel, allows separation of the polarized and unpolarized portions of a partially polarized signal.

Polarization imaging was evaluated as a method to measure the absorbed dose to water from the Cherenkov signal [8, 9, 10, 11]. The method of Cloutier et al demonstrated the ability to isolate the anisotropic signal contribution, which could then be corrected using known angular Cherenkov distributions [9]. This correction process can be represented mathematically as:

$$D(x, y) \propto \mathcal{C}_\theta(x, y) \cdot \mathcal{C}_\phi(x, y) \cdot I_{pol}(x, y). \tag{4}$$

Monte Carlo simulations were employed to generate the polar (θ) and azimuthal (ϕ) angular correction distributions $\mathcal{C}_\theta(x, y)$ and $\mathcal{C}_\phi(x, y)$, respectively. Profiles and depth dose measurements were conducted using photon (6 MV and 18 MV) and electron (6 MeV and 18 MeV) treatment beams. The overall polarized corrected Cherenkov signals exhibited mean differences within 3% when compared to radiochromic films under all irradiation conditions. For profiles, corrected polarized Cherenkov measurement showed better agreement with expected doses on the principal axis. In the beam umbra, an under-response was observed, attributed to the predominance of electron deposition below the Cherenkov production threshold ($E_{min} = 260$ keV). Furthermore, it was demonstrated that the same angular correction function could be applied throughout the volume for photon measurements, as these distributions (both polar and azimuthal) did not vary with depth or off-axis positions. However, electron beam polar angular distributions varied with depth, necessitating comprehensive Monte Carlo simulations. The system presented in that work [9] was limited to 2D dose measurements, summing signals over the water tank thickness. Subsequently, Toyonaga et al. also presented an approach using two sets of analyzers to correct the signal [11]. This method was applied to photon measurements only and in a specific geometry [8, 9, 10, 11]. These results all showcased the potential of polarization imaging for developing new, truly water-equivalent radiation detectors. Hence, this manuscript aims to presents new developments regarding the application of polarization imaging using the Cherenkov signal for measuring ultra-high dose rate electron beams.

2 Material and Methods

A detector suitable for quality assurance of ultra-high dose rate (UHDR) intra-operative electron beams was developed. In this approach, the polarized Cherenkov signal serves as a surrogate quality assurance device, prioritizing precision over accuracy, eliminating the need for Monte Carlo correction factors. The Mobetron intraoperative electron beam linear accelerator was characterized using this approach.

The Cherenkov signal emitted in a water tank was captured using a CCD camera (414EX; Atik Cameras, Norwich, UK) coupled to a rotating polarizer (figure 1). The wide dynamic range of the CCD camera facilitated measurements for both conventional and UHDR doses. Images were acquired from four transmission axes (0°, 45°, 90°, 135°), with an integration time of 5s, and the polarized portion of the signal was isolated using the previously described method using Malus law. For every measurement, either 50 monitor units (approximately 90 pulses) were administered in conventional mode or 10 pulses in UHDR mode. Percent depth Cherenkov signal (PDC) were measured for different collimator sizes (⌀:6 and 3cm), nominal beam energies (6 and 9 MeV) and dose rates (UHDR, conventional). The linearity of the signal with the pulse width and number of pulses was also validated. The linearity curves were determined based on the signal measured at $d_{\max}$ on the 6 MeV beam PDC.

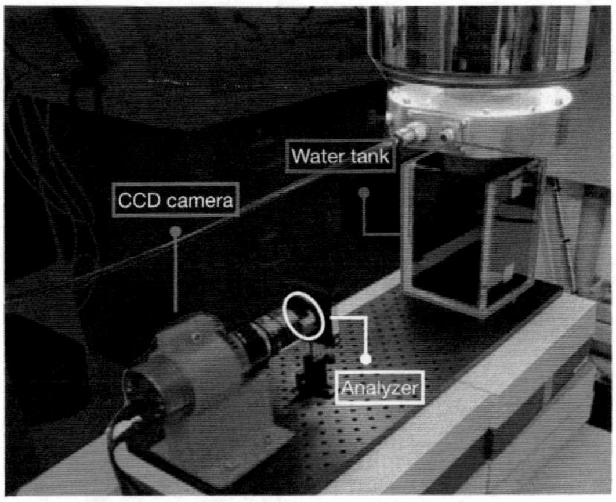

Figure 1: Measurement setup comprising a CCD camera coupled to a rotating analyzer imaging a water tank (width = 10.9 cm, depth = 10.9 cm, height = 11.3 cm). The collimator applicator is normally incident to the water surface.

3 Results

Reproducibility was measured by repeating each set of measurements three times, which resulted in variations below 0.1%. Percent depth Cherenkov signal for each measurement are presented in Figure 2, and Cherenkov beam quality, defined as C_{50} was measured with a precision of 0.05 cm. The Cherenkov beam quality did not vary significantly with dose rate nor field size. The signal exhibited a high degree of linearity with respect to both the number of pulses and the pulse width, with a coefficient of determination (R^2) exceeding 0.99, as illustrated in Figure 3. Low signal issues related to Cherenkov measurements were not encountered in UHDR mode, as the pulses delivered, on average, 320 times more signal than conventional pulses.

4 Discussion

The investigation of polarization imaging has opened new possibilities for the application of Cherenkov signal in radiation therapy dosimetry. The technique has been successfully applied to directly measure the absorbed dose to water using the Cherenkov signal, demonstrating high precision and agreement with radiochromic films when correcting for its angular distribution. Ultra-high dose rates present challenges for conventional detectors such as ionization chambers, including saturation, radiation damage and recombination issues [12]. Hence, novel approaches for routine quality assurance are required to enable the clinical translation of ultra-high dose rate electron radiotherapy. The results presented, demonstrating signal linearity, repeatability, and the possibility of beam quality monitoring, hold promise for the advancement of real-time quality assurance devices capable of surrogate monitoring of ultra-high dose rate radiation beams. Further work is required to properly assess the sensitivity of C_{50} in detecting beam quality changes, because the Cherenkov signal is only produced by secondary electrons with energies above a production threshold.

Other luminescent detectors, using scintillation rather than the Cherenkov signal, have also shown interesting performance for electron-beam ultra-high dose rate measurements [13, 14]. Radioluminescent detectors are distinguished from charge and chemical detectors by combining high spatial and temporal resolution while allowing measurements over a wide range of dose rates. [15]. Yet, plastic scintillation detectors have only sampled a single point in ultra-high dose rate conditions and are susceptible to radiation damage, which needs to be quantified.

Journal of Physics: Conference Series **2799** (2024) 012015 doi:10.1088/1742-6596/2799/1/012015

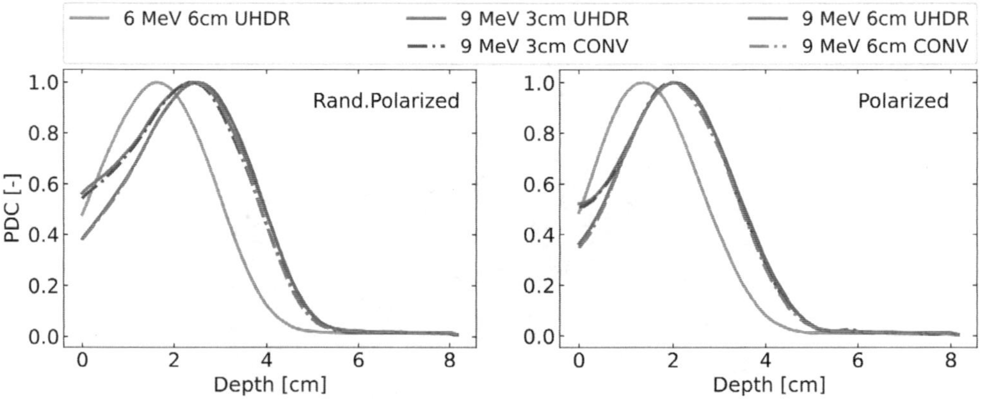

Figure 2: Percent depth Cherenkov signal (PDC) curves measured from different electron beam energy, dose rates and field sizes. Results from the randomly polarized portion of the signal are presented on the left panel whereas the polarized portion of the signal is presented on the right panel.

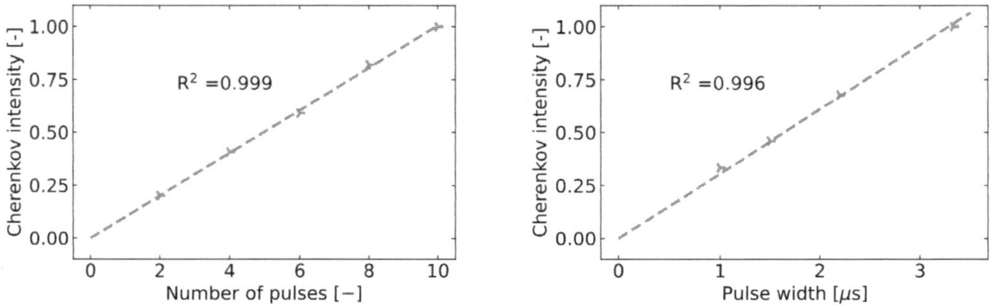

Figure 3: Linearity of the Cherenkov signal with the number of pulses and the effective pulse width. Curves results from the 6 MeV beam measurements.

5 Conclusion

A novel approach was explored for quality assurance of ultra-high dose rate intra-operative electron beams. In this approach, the polarized Cherenkov signal serves as a surrogate quality assurance device, providing precision measurements without the need for angular corrections. The results showed high linearity of the Cherenkov signal with the number of pulses and pulse width, making it a promising candidate for real-time quality assurance devices for both conventional and ultra-high dose rate radiation beams.

References

[1] M. F. L'Annunziata, "Chapter 6 - cherenkov counting," in *Handbook of Radioactivity Analysis (Third Edition)* (M. F. L'Annunziata, ed.), pp. 935–1019, Amsterdam: Academic Press, third edition ed., 2012.

[2] A.-M. Frelin, J.-M. Fontbonne, G. Ban, J. Colin, M. Labalme, A. Batalla, A. Isambert, A. Vela, and T. Leroux, "Spectral discrimination of Čerenkov radiation in scintillating dosimeters," *Medical Physics*, vol. 32, no. 9, pp. 3000–3006, 2005.

[3] F. Therriault-Proulx, L. Beaulieu, L. Archambault, and S. Beddar, "On the nature of the light produced within PMMA optical light guides in scintillation fiber-optic dosimetry," *Physics in Medicine and Biology*, vol. 58, no. 7, pp. 2073–2084, 2013.

[4] A. K. Glaser, S. C. Davis, W. H. A. Voigt, R. Zhang, B. W. Pogue, and D. J. Gladstone, "Projection imaging of photon beams using cerenkov-excited fluorescence," *Physics in Medicine & Biology*, vol. 58, no. 3, p. 601, 2013.

[5] E. Jean, M.-E. Delage, and L. Beaulieu, "Investigation of the quinine sulfate dihydrate spectral properties and its effects on cherenkov dosimetry," *Physics in Medicine and Biology*, vol. 64, no. 15, p. 155019, 2019.

[6] Y. Zlateva, B. R. Muir, I. El Naqa, and J. P. Seuntjens, "Cherenkov emission based external radiotherapy dosimetry: I. formalism and feasibility," *Medical Physics*, vol. 46, no. 5, pp. 2370–2382, 2019.

[7] Y. Zlateva, B. R. Muir, J. P. Seuntjens, and I. El Naqa, "Cherenkov emission based external radiotherapy dosimetry: II. electron beam quality specification and uncertainties," *Medical Physics*, vol. 46, no. 5, pp. 2383–2393, 2019.

[8] E. Cloutier, L. Archambault, and L. Beaulieu, "Accurate dose measurements using cherenkov emission polarization imaging," *Medical Physics*, vol. 49, pp. 5417–5422, 2022.

[9] E. Cloutier, L. Beaulieu, and L. Archambault, "Direct in-water radiation dose measurements using cherenkov emission corrected signals from polarization imaging for a clinical radiotherapy application," *Scientific Reports*, vol. 12, no. 9608, 2022.

[10] S. Yamamoto, T. Yabe, T. Akagi, and Y. Hirano, "Imaging of polarized components of cerenkov light and luminescence of water during carbon ion irradiation," *Medical Physics*, vol. 48, pp. 427–433, 2020.

[11] C. Toyonaga, S. Yamamoto, T. Yabe, K. Okudaira, K. Yogo, Y. Hirano, and J. Kataoka, "Technical note: Correcting angular dependencies using non-polarized components of cherenkov light in water during high-energy x-ray irradiation," *Medical Physics*, vol. 49, no. 8, pp. 5409–5416, 2022.

[12] F. Romano, C. Bailat, P. G. Jorge, M. L. F. Lerch, and A. Darafsheh, "Ultra-high dose rate dosimetry: Challenges and opportunities for flash radiation therapy," *Medical Physics*, vol. 49, no. 7, pp. 4912–4932, 2022.

[13] Y. Poirier, J. Xu, S. Mossahebi, F. Therriault-Proulx, and A. Sawant, "Technical note: Characterization and practical applications of a novel plastic scintillator for online dosimetry for an ultrahigh dose rate (flash)," *Medical Physics*, vol. 49, no. 7, pp. 4682–4692, 2022.

[14] E. Ciarrocchi, E. Ravera, A. Cavalieri, M. Celentano, D. Del Sarto, F. Di Martino, S. Linsalata, M. Massa, L. Masturzo, A. Moggi, M. Morrocchi, J. Pensavalle, and M. Bisogni, "Plastic scintillator-based dosimeters for ultra-high dose rate (uhdr) electron radiotherapy," *Physica Medica*, vol. 121, p. 103360, 2024.

[15] M. R. Ashraf, M. Rahman, R. Zhang, B. B. Williams, D. J. Gladstone, B. W. Pogue, and P. Bruza, "Dosimetry for flash radiotherapy: A review of tools and the role of radioluminescence and cherenkov emission," *Frontiers in Physics*, vol. 8, p. 328, 2020.

Journal of Physics: Conference Series **2799** (2024) 012016 doi:10.1088/1742-6596/2799/1/012016

Real-time Plastic Scintillation Dosimetry of Ultra-High Dose Rate Very High Energy Electrons (VHEE) at CERN CLEAR Facility

Cloé Giguère[1,2*], Alexander Hart[3], Joseph Bateman[4], Pierre Korysko[4,5], Wilfrid Farabolini[5], Yoan LeChasseur[6], Magdalena Bazalova-Carter[3], Luc Beaulieu[1,2]

[1] Département de Physique, de Génie Physique et d'Optique et Centre de Recherche sur le Cancer, Université Laval, Québec, QC G1V 0A6, Canada.
[2] Département de Radio-Oncologie et Axe Oncologie du CRCHU de Québec, CHU de Québec, Université Laval, Québec, QC G1V 0A6, Canada.
[3] Department of Physics and Astronomy, University of Victoria, Victoria, BC V8P 5C2, Canada.
[4] Department of Physics, University of Oxford, OX1 3AZ Oxford, U.K.
[5] CERN, 1211 Geneva, Switzerland
[6] Medscint, Québec, QC, Canada

*E-mail: clgig68@ulaval.ca

Abstract. Very High Energy Electrons (VHEE) that can theoretically treat deep-seated tumours and be delivered at ultra-high dose rates (UHDR) could be the solution to translate FLASH radiotherapy into the clinic. Standard dosimeters have limited application in those extreme conditions, but dose-rate independent and fast-response plastic scintillation detectors (PSDs) are a potential alternative to overcome this. In this work, response of a 4-channel PSD to the 200 MeV VHEE UHDR beam delivered with doses and dose rates in pulse up to 90 Gy and 4.6×10^9 Gy/s, respectively, at the CLEAR facility in CERN was characterized, using the Hyperscint RP200 platform. Scintillation light linearity with dose was observed for three scintillators from ~5-50 Gy, while clear fiber output was linear up to 90 Gy. While linearity on this dose range was conserved even after radiation damage by exposure to 37.2 kGy total accumulated dose, light output significantly decreased. This work proves the potential of plastic scintillators for real-time dosimetry of UHDR VHEE beams.

1. Introduction

Conventional external radiotherapy typically uses a beam of ionizing radiation, either electrons or photons, produced by a medical linear accelerator (linac) to kill cancer cells. Decades of research have allowed modern radiotherapy treatments to accentuate the differential response of healthy tissues and tumours to radiation. However, doses delivered are still limited by normal tissue tolerance, as severe side effects must be avoided. As shown in recent years, delivering radiation at ultra-high dose rates (UHDR), with single fraction doses in sub second treatment times, can trigger a normal tissue sparing effect while maintaining toxicity to tumour cells [1-3]. This effect, termed the FLASH effect, has been demonstrated in many animal models with different particle modalities and beam characteristics [4]. Low energy electrons are widely used right now to produce UHDR beams, but their ability to treat tumours is limited because of their low penetration in the body. Very High Energy Electrons (VHEE) beams (>100 MeV) offer an interesting avenue for external radiotherapy due to their superior penetration capabilities and favourable dose distribution that allows treatment of deep-seated tumours. They are less sensitive to inhomogeneities than protons and have the potential to treat tumours as effectively as or even better than photon beams [5], [6]. High-energy electrons, deliverable at high dose rates, are thus promising candidates for translating FLASH therapy into clinical practice.

UHDR VHEE beams pose new challenges for accurately measuring dose under conditions that are far removed from reference clinical measurements. Radiochromic films, extensively used in the clinic for 2D dose measurements, have excellent dose rate independence, but they do not allow real-time measurements [7]. The ionization chamber, the reference dosimeter in radiotherapy, suffers from significant charge recombination at high dose rates, which causes change in sensitivity, non-linearity and saturation [8]. A possible solution to overcome these limitations lies in the development of water equivalent plastic scintillation detectors (PSDs) that can be read in real-time. Comprised of a scintillating material emitting light when irradiated, PSDs proved to be promising detectors for dosimetry at ultra-high dose rates in UHDR low energy electrons and photons [9], [10], [11]. A recent study by Hart *et al.* demonstrated their utility for real-time dosimetry of 200 MeV UHDR electrons [12]. This work aims to further investigate the output linearity, radiation damage and recovery of PSDs when exposed to the UHDR VHEE beam delivered at the CLEAR CERN facility.

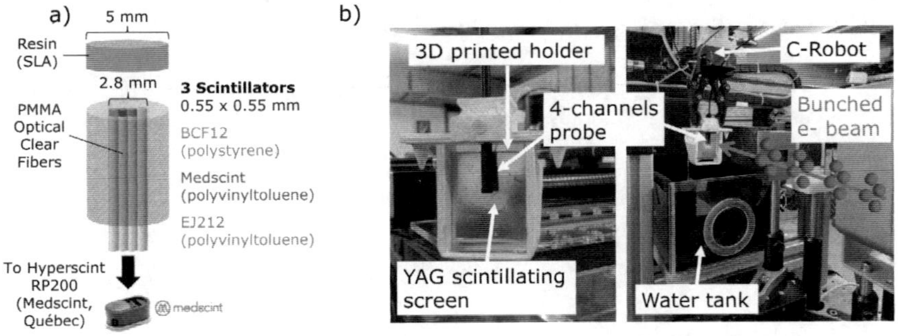

Figure 1. a) 4-channel PSD design. b) Experimental set-up in the in-air test area of the CLEAR beam line.

2. Materials and Methods

A 4-channel PSD (Figure 1a) was designed with polystyrene based BCF12, polyvinyltoluene (PVT) based EJ-212, PVT based proprietary Medscint scintillators and clear plastic fiber for stem effect light (Cherenkov and fluorescence) removal. The probe, installed on a 3D printed holder held by the C-Robot, was centred on the gaussian beam visually using the reflection of a YAG scintillating screen (Figure 1b). An integrating current transformer (ICT) recorded delivered charges per pulse, that were then converted to dose with Gafchromic MD-V3 films measurements performed on each day of experiments. The channels were connected to the Hyperscint RP200 platform to measure scintillation light output to dose pulses from ~5 to 90 Gy. Dose rates in pulse varied throughout but ranged from 2.2×10^{8} to 4.6×10^{9} Gy/s. Between measurements, the PSD was damaged with irradiations of ~5 kGy delivered with about 12 Gy pulses at 3.33 Hz for dose rates of ~40 Gy/s in average and ~1.4×10^{9} Gy/s in pulse. In total, 37.2 kGy was delivered to the probe. Linearity of response to dose was computed by plotting the area under the curve from the measured output spectra according to dose per pulse, after subtracting background and stem effect. Slopes of linear measurements, extracted from linear regressions, were taken as light output. Linearity limits were determined using visual inspection, the coefficients of determination (R^2) and residuals of the linear regressions. In between the linearity measurements, scintillators were allowed to rest for a given period of time, then re-irradiated to assess short-term recovery of light output and spectral changes. To evaluate long-term recovery, output and spectra measurements were conducted respectively with the 6 MV beam and the kV source of a TrueBeam clinical Linac at time points ranging from 24 to 237 days after radiation damage and compared to the same measurements done before CLEAR irradiations.

3. Results and Discussion

Figure 2 shows the scintillator and clear fiber signal according to dose per pulse for increasing total accumulated dose. Irregularities in the curves are likely caused by variations in beam size and positioning. For scintillators, integrated light signal is linear to dose per pulse until about 50 Gy/pulse (~8×10^8 Gy/s in pulse), while clear fiber output stayed linear over the whole dose and dose rate range studied. A saturation of scintillator response is observed for doses over 50 Gy/pulse, which was not caused by a saturation of the photodetector system. Dose rates varied non linearly with dose per pulse and higher dose rates were sometimes used for smaller doses due to combination of number of bunches and pulse length. Saturation cause is thus difficult to isolate, but could be caused by a decrease of scintillation efficiency at high dose per pulse due to a quenching effect normally seen for high LET radiation.

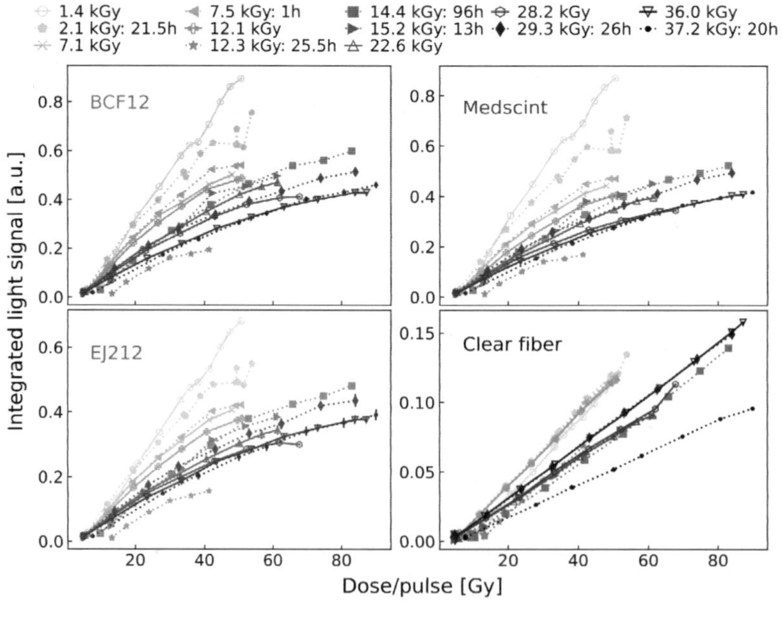

Figure 2. Integrated light signal (area under the curve) of scintillator and clear fiber spectra measured for different dose per pulse, at increasing total accumulated dose in the probe. Mean over 3 single pulses is shown. Full symbols represent short-term recovery measurements, with associated rest times specified in the legend. Errors in x, ranging from about ± 1 to 11 Gy and not shown for visual clarity, were calculated from standard deviation of dose pulses, uncertainty in dose calibration and uncertainty in probe positioning.

Light output (linear slopes) significantly decreased with accumulated dose in the probe, but stayed linear to dose for the ranges previously discussed. PSDs could thus still be used for relative dose measurements even when damaged by radiation. Mean losses of light yield of <1.85 %/kGy were measured for scintillators and clear fiber. Short-term output recovery reported at 1h, 96h, 13h and 26h following irradiation suggests that recovery depends on inter-irradiation time interval but also on accumulated dose. Scintillation and clear fiber spectra were shifted towards higher (yellower) wavelengths with dose, which is consistent with the yellow discoloration that normally appears in irradiated plastic scintillators and fibers [13].

Long-term recovery of light output loss was significant, but partial, and appears to stabilize after 39 days for BCF12, between 65 and 150 days for Medscint and EJ212 and after 32 days for clear fiber.

Conclusion

This study demonstrates that plastic scintillators and clear fibers respond linearly with dose in VHEE UHDR conditions for a clinically relevant dose range, even after exposure to 37.2 kGy. While scintillators lost linearity at about 50 Gy/pulse, clear fibers performed better with no loss of linearity on the whole dose range measured. Significant loss of light output <1.85 %/kGy and spectral shift with radiation damage was observed, as well as short-term (<100h) recovery. After less than 150 days of rest, output partially recovered. These results are a critical step toward the use of plastic scintillators for real-time dosimetry of UHDR VHEE beams. Future work will aim to use less variable FLASH beams to verify some of the conclusions of this work with less dose uncertainty. Recovery processes with scintillator compositions designed for increased radiation hardness will also be explored, as well as response saturation at high doses per pulse.

References

[1] M.-C. Vozenin, J. H. Hendry, and C. L. Limoli, "Biological Benefits of Ultra-high Dose Rate FLASH Radiotherapy: Sleeping Beauty Awoken," *Clin. Oncol.*, vol. 31, no. 7, pp. 407–415, Jul. 2019, doi: 10.1016/j.clon.2019.04.001.

[2] V. Favaudon *et al.*, "Ultrahigh dose-rate FLASH irradiation increases the differential response between normal and tumor tissue in mice," *Sci. Transl. Med.*, vol. 6, no. 245, p. 245ra93, Jul. 2014, doi: 10.1126/scitranslmed.3008973.

[3] P. Montay-Gruel *et al.*, "Irradiation in a flash: Unique sparing of memory in mice after whole brain irradiation with dose rates above 100 Gy/s," *Radiother. Oncol.*, vol. 124, no. 3, pp. 365–369, Sep. 2017, doi: 10.1016/j.radonc.2017.05.003.

[4] N. Esplen, M. S. Mendonca, and M. Bazalova-Carter, "Physics and biology of ultrahigh dose-rate (FLASH) radiotherapy: a topical review," *Phys. Med. Biol.*, vol. 65, no. 23, p. 23TR03, Dec. 2020, doi: 10.1088/1361-6560/abaa28.

[5] M. Bazalova-Carter *et al.*, "Treatment planning for radiotherapy with very high-energy electron beams and comparison of VHEE and VMAT plans," *Med. Phys.*, vol. 42, no. 5, pp. 2615–2625, 2015, doi: 10.1118/1.4918923.

[6] L. Whitmore, R. I. Mackay, M. van Herk, J. K. Jones, and R. M. Jones, "Focused VHEE (very high energy electron) beams and dose delivery for radiotherapy applications," *Sci. Rep.*, vol. 11, p. 14013, Jul. 2021, doi: 10.1038/s41598-021-93276-8.

[7] M. R. Ashraf *et al.*, "Dosimetry for FLASH Radiotherapy: A Review of Tools and the Role of Radioluminescence and Cherenkov Emission," *Front. Phys.*, vol. 8, p. 328, Aug. 2020, doi: 10.3389/fphy.2020.00328.

[8] M. McManus *et al.*, "The challenge of ionisation chamber dosimetry in ultra-short pulsed high dose-rate Very High Energy Electron beams," *Sci. Rep.*, vol. 10, no. 1, Art. no. 1, Jun. 2020, doi: 10.1038/s41598-020-65819-y.

[9] J. Archer, E. Li, J. Davis, M. Cameron, A. Rosenfeld, and M. Lerch, "High spatial resolution scintillator dosimetry of synchrotron microbeams," *Sci. Rep.*, vol. 9, no. 1, p. 6873, May 2019, doi: 10.1038/s41598-019-43349-6.

[10] Y. Poirier, J. Xu, S. Mozzsahebi, F. Therriault-Proulx, and A. Sawant, "Technical note: Characterization and practical applications of a novel plastic scintillator for online dosimetry for an ultrahigh dose rate (FLASH)," *Med. Phys.*, vol. 49, no. 7, pp. 4682–4692, 2022, doi: 10.1002/mp.15671.

[11] A. Hart *et al.*, "Lead-doped scintillator dosimeters for detection of ultrahigh dose-rate x-rays," *Phys. Med. Biol.*, vol. 67, no. 10, p. 105007, May 2022, doi: 10.1088/1361-6560/ac69a5.

[12] A. Hart *et al.*, "Plastic Scintillator Dosimetry of Ultrahigh Dose-Rate 200 MeV Electrons at CLEAR," *IEEE Sens. J.*, vol. 24, no. 9, pp. 14229–14237, May 2024, doi: 10.1109/JSEN.2024.3353190.

[13] C. Zorn, "A pedestrian's guide to radiation damage in plastic scintillators," *Nucl. Phys. B - Proc. Suppl.*, vol. 32, pp. 377–383, May 1993, doi: 10.1016/0920-5632(93)90049-C.

Journal of Physics: Conference Series **2799** (2024) 012017 doi:10.1088/1742-6596/2799/1/012017

On the use of Al_2O_3:C and plastic scintillators for small-field dosimetry in MR-linacs

Claus E. Andersen*, Mads F. Klavsen and Christina Ankjærgaard

Department of Health Technology, Technical University of Denmark, Roskilde, Denmark

*E-mail: clan@dtu.dk

Abstract. Fiber coupled luminescence detectors based on organic plastic scintillators or inorganic materials such as Al_2O_3:C are suitable for dosimetry in strong magnetic fields as found in MR-linacs. The main finding of this work was that field-output factor measurements at the maximum dose point are less variable than if measured at the central beam axis (CAX). In contrast to plastic scintillators, Al_2O_3:C was found to have a yield of radioluminescence per dose that (within uncertainty) is independent of the magnetic field strength.

1. Introduction

Small-field dosimetry in megavoltage photon beams is a significant measurement challenge for air-filled ionization chambers due to the absence of charge-particle equilibrium. In contrast, plastic scintillators with a density and an atomic composition close to water can provide near perturbation-free small-field dosimetry in a water phantom for MV photon beams. The benefit of such solid-state detectors increases even further for measurements in MR-linacs where the presence of a strong magnetic field (e.g. 0.35 T or 1.5 T) can lead to additional perturbation of the electron fluence if the detector density deviates from that of water [1-3]. Another significant feature of fiber-coupled plastic scintillators is that they do not the disturb MR imaging [4,5].

The primary objective of this work was to identify the best position for field-output factor measurements in the presence of a strong magnetic field. Specifically, we compared the closeness of agreement of field output factors in an electromagnet set-up at different B-fields obtained at the maximum dose point compared with a point on the central axis of the beam. Such measurements can be performed with plastic scintillator as no significant correction are needed [6] except with regards to the stem removal and volume averaging. A secondary objective was to study the B-field induced changes of the light yield per dose for fiber-coupled Al_2O_3:C radioluminescence detectors. This part of the study was driven by the wish to make a comparison against recently published light-yield data for BCF-60 organic plastic scintillators [7], that were found to be strongly affected by the magnetic field even below 10 mT. Furthermore, fiber-coupled Al_2O_3:C radioluminescence is interesting, since the stem signal can be removed completely in pulsed beams by time separation [8,9], and since this measurement technique provides output factors similar to plastic scintillators and alanine [10].

2. Material and Methods

Experiments were carried out using a strong electromagnet (max 0.7 T with 10 cm pole shoe gap) and horizontal 6 MV FFF beams from medical linear accelerator (Truebeam, Varian Medical

Journal of Physics: Conference Series **2799** (2024) 012017 doi:10.1088/1742-6596/2799/1/012017

Systems, USA). Fiber-coupled dosimetry was performed at 7 cm depth of water in a small water phantom (width, height, depth: 10 cm × 19 cm × 21 cm) placed within the electromagnet at an extended distance of 181.6 cm from the linac virtual source. Measurements were carried out at different field sizes using both BCF-60 scintillating fiber (Saint-Gobain, USA) and fiber-coupled Al_2O_3:C crystals (Landauer, Stillwater Crystal Growth Division, USA). All measurements presented here were with the fiber cable and the detector in vertical orientation. The beam and the magnetic field were perpendicular to each other in the horizontal plane. FPGA-based instrumentations (ME30 and ME40) developed at the Technical University of Denmark were used for the read out. The ME30 instrument allowed for time separation of the radioluminescence (RL) from Al_2O_3:C and the stem signal (primarily Cerenkov light produced in the optical fiber cable) following the saturated RL measurement protocol developed previously [8]. Likewise, the ME40-system reduced the influence of the stem effect on the scintillator signal by a two-channel chromatic removal technique [11,12]. Monte-Carlo modelling with TOPAS version 3.9 [13] was carried out to derive correction factors for the ratio between the dose to Al_2O_3:C and water. The light yield measurements followed the same procedure as reported elsewhere [7] and the dose to water at the position of the measurement was established in the same way for each magnetic field strength.

3. Results

Figure 1 shows field output factors obtained with the BCF-60 scintillator system for nominal field sizes from 0.5 cm x 0.5 cm to 5 cm x 5 cm at the-iso center (i.e. from about 0.9 cm x 0.9 cm to 9 cm x 9 cm at the point of measurement) for 0 T, +/-0.35 T, and +/- 0.7 T magnetic fields. The field output factors measured at the maximum dose point were found to be essentially independent of the B-field. This was not the case for the measurements at the central axis of the beam. The results of the light-yield measurements for Al_2O_3:C are shown in Figure 2 for a nominal 5 cm x 5 cm field. We obtained the dose to Al_2O_3:C using the Monte-Carlo correction factors in Table 1 and the assumption (tested in [7]) that the dose to water was the independent of the B-field. Within the limitations of the experiment (0.5% relative uncertainty, k=1), the magnetic field was not found to affect the Al_2O_3:C radioluminescence yield per dose in the tested range from 0 T to 0.7 T.

4. Discussion

Clearly, measurements of field output factors need to be performed under conditions relevant for the end use and that may require measurements at the central beam axis. However, measurements at the maximum dose point have an advantage since they are essentially independent of the B-field at least under the conditions tested in this study [14]. The absence of any significant influence of the magnetic field on the radioluminescence from Al_2O_3:C is interesting since the light yield per dose for the BCF-60 scintillator (green component) was previously found to be strongly influenced by the magnetic field with an increase in light yield per dose of (4.5±0.9)% (k=1) from 0 T to 0.7 T [7]. Since the stem signal can be fully removed in fiber-coupled Al_2O_3:C dosimetry using time separation [9], Al_2O_3:C provides a useful dosimetry tool for measurements in strong magnetic fields. From a dosimetric point of view [15], plastic scintillators, however, remain to be the preferred detector type given Al_2O_3:C's density (4 g/cm^3) and effective atomic number (11.3). Previous measurements [16] using the optically stimulated luminescence signal have indicated that Al_2O_3:C has a ~1% reduction in signal output at 1 T.

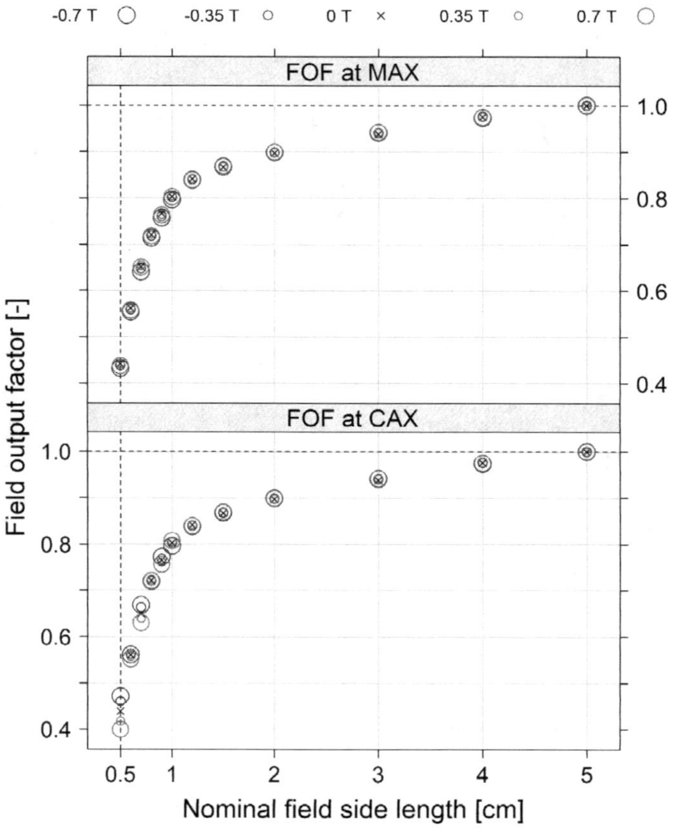

Figure 1. Field output factors (FOFs) measured with the BCF-60 scintillator at the maximum dose point (top panel) or at the central axis (CAX, bottom panel) of the 6 MV FFF beam in the water phantom positioned in the electromagnet. The measurements were carried out at different magnetic field strengths and nominal field sizes as indicated.

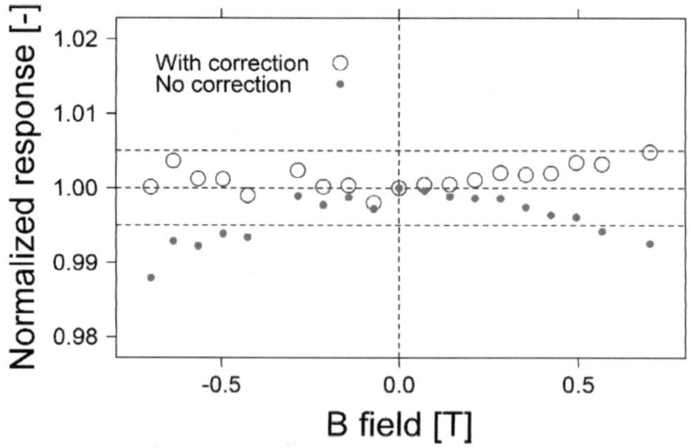

Figure 2. Response (light yield per dose to Al2O3:C) vs. magnetic field strength for the 2 mm Al2O3:C crystal during irradiations in water for a broad 6 MV FFF beam. The open circle data have been corrected for the influence of the magnetic field on the ratio between dose to water (established with a reference detector) and Al2O3:C (Table 1). Dashed lines show the 0.5% deviation from unity.

Table 1. Monte-Carlo computations of the influence of magnetic field on the ratio between dose to the Al2O3:C crystal and the dose to a water voxel of an identical size (normalized at 0 T). The uncertainties in the table are standard uncertainties (k=1) only representing the statistical nature of the Monte Carlo technique. The crystal used for the measurements was 2 mm long, the table includes a supplementary computation for a 10 mm crystal.

Crystal	Magnetic field strength			
length	0 T	0.2 T	0.4 T	0.7 T
2 mm	1.0000	0.9979 ± 0.0021	0.9969 ± 0.0023	0.9875 ± 0.0021
10 mm	1.0000	0.9978 ± 0.0011	0.9963 ± 0.0009	0.9860 ± 0.0010

5. Conclusion

This work demonstrated the use for fiber-coupled luminescence detectors in strong magnetic field (up to 0.7 T). It was found that field output factor measurements at the maximum dose point were essentially independent of the magnetic field strength. This implies that the uncertainty of detector correction factors can be minimized in an electromagnet set-up if measured at the maximum dose point. Al_2O_3:C may not have the same high degree of water equivalence as organic plastic scintillators [15] and the response is expected to depend on the B-field as demonstrated by the Monte-Carlo computations in Table 1. However, Al_2O_3:C benefits from having a radioluminescence yield per dose to Al_2O_3:C that is independent of the magnetic field from 0 to 0.7 T, and the stem signal can be completely removed using time separation technique.

Acknowledgments

This work was carried out within the framework of the project "Traceable dosimetry for small fields in MR-guided radiotherapy" funded by the EMPIR programme (19NRM01) co-financed by the Participating States and from the European Union's Horizon 2020 Research and Innovation Programme.

References

[1] Uijtewaal P et al. 2023 *Phys. Med. Biol.* **68** 4 04NT01.
[2] Oolbekkink S et al. 2024. *J. Appl Clin Med Phys.* **25** 1 e14180.
[3] Madden L et al. 2020 *Physica Medica* **73** 111–116.
[4] Klavsen M F et al. 2022 *Radiat. Meas.* **154** 106759.
[5] Klavsen M F et al. 2023 *Biomed Phys Eng Express* **9** 6 065001.
[6] Santurio G V & Andersen C E 2019 *Radiat. Meas.* **129** 106200.
[7] Andersen C E 2024 *Biomed. Phys. Eng. Express* **10** 1 015016.
[8] Andersen C E et al. 2011 *Radiat. Meas.* **46** 10 1090-1098.
[9] Buranurak S & Andersen C E 2016 *Radiat. Meas.* **93** 46-54.
[10] Azangwe G et al. 2014 *Med. Phys.* **41** 7 072103.
[11] Beierholm A R et al. 2015. *Radiat. Meas.* **82** 129-137.
[12] Frelin, A M et al. 2006 *IEEE Trans. on Nucl. Sci.* **53** 1113-1116.
[13] Perl J et al. 2012 *Med. Phys.* **39** 11 6818-6837.
[14] Klavsen M F et al. 2024 *Radiat. Meas.* **175** 107163.
[15] Beierholm A R et al. 2008 *Radiat. Meas.* **43** 2-6 898–903.
[16] Spindeldreier C K et al. 2017 *Radiat. Meas.* **106** 346-351.

Journal of Physics: Conference Series **2799** (2024) 012018 doi:10.1088/1742-6596/2799/1/012018

Dosimetric Evaluation of an Adaptive Planning Strategy for Implant Shift in HDR Prostate Brachytherapy

Jordan Wallace[1*], Maximilian Hanlon[1,2], Ryan L Smith[1,2,3], and Rick Franich[1,2]

[1] School of Science, RMIT University, Melbourne VIC, Australia
[2] Alfred Health Radiation Oncology, Melbourne VIC, Australia
[3] Central Clinical School, Monash University, Melbourne VIC, Australia

[*]E-mail: jordan.wallace@rmit.edu.au

Abstract. The occurrence of significant implant shifts between CT simulation planning and first fraction delivery is a frequent challenge for HDR prostate brachytherapy treatment. Here, we present evidence supporting the viability of a first-order adaptation strategy as a viable alternative to the conventional practice of re-scanning and re-planning. Our findings underscore the potential of this approach to enhance workflow efficiency in HDR prostate brachytherapy.

1. Introduction

Validation of correct clinical delivery of high dose rate (HDR) prostate brachytherapy relies on verifying that the implant geometry and patient anatomy are concordant with the plan. Implant geometrical changes that may have occurred could alter the dose delivered to the tumour but can readily be assessed via pre-treatment imaging (1). Gross implant shifts can trigger a need for re-planning, but this is a workload intensive procedure.

The purpose of this work is to investigate the feasibility of using pre-treatment imaging of the HDR prostate implant as a basis to perform a first-order online adaptive treatment planning strategy. The proposed approach uses the implant shift assessment to adapt the original treatment plan while preserving the dosimetric planning objectives as an alternative to a complete re-plan.

2. Materials and Methods

2.1 Treatment Procedure

HDR prostate brachytherapy at the Alfred Health Radiation Oncology (AHRO) is performed as a conformal boost to the external beam radiotherapy (EBRT) schedule. During this study, the boost is delivered in two fractions of 8.5 Gy, with one insertion, delivered in successive days. CT imaging and planning is performed at each treatment fraction. The patient is transferred to the treatment bunker and moved onto the couch before pre-treatment imaging followed by treatment delivery.

2.2 Pre-treatment Imaging Procedure at AHRO

The treatment couch is a customised surgical couch, modified with a flat panel detector embedded into the frame beneath the patient's pelvis. It is used to perform pre-treatment

Journal of Physics: Conference Series **2799** (2024) 012018 doi:10.1088/1742-6596/2799/1/012018

Anterior-Posterior (AP) imaging to assess catheter displacement. Imaging is performed immediately prior to treatment delivery, with radio-opaque marker wires inserted into 5 catheters providing visualisation in the AP radiograph (1,2). Catheter paths are compared to those defined in the treatment planning system (TPS) as seen in Figure 1. Superior-Inferior (SI) catheter tip displacements are measured as indicated by the red arrows in Figure 1. If the average measured inferior displacement of the 5 catheter tips is less than 7 mm, it is within our acceptable tolerance and treatment proceeds. If catheter displacement exceeds the 7 mm threshold, the patient is re-CT scanned and a new plan is created, as shown by the workflow in Figure 2. The 7 mm threshold was determined from a retrospective analysis of plan robustness at our centre.

2.3 Patients

In this study, all data analysed was for patients with measured displacements exceeding the 7 mm threshold, and so were re-scanned and re-planned. Pre-treatment imaging and source tracking have been routinely recorded for all patients treated with HDR brachytherapy at AHRO since 2014. Plans for 47 previous patients were assessed for catheter displacement. Of those patients, 23 exceeded the clinical threshold for treatment and were subsequently replanned. In this work, 9 patients were identified to have implant displacement exceeding the 7 mm threshold and were subsequently re-scanned for re-planning without having a manual advancement of the implant. The updated anatomical and catheter positioning data allowed for this assessment of a first order plan adaption method.

2.4 Dosimetry of Adapted Plan

An adaption of the original treatment plan was created for each patient by applying an offset to all dwells in the clinical plan equal to the average catheter tip displacement measured during

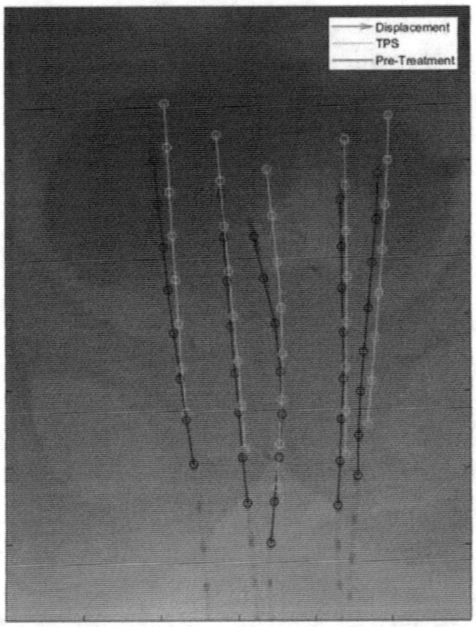

Figure 1. A pre-treatment A-P image acquired with radio-opaque marker wires inserted into 5 catheters. The marker wire tips (blue) are compared to the planned tip positions (green) for SI shift analysis.

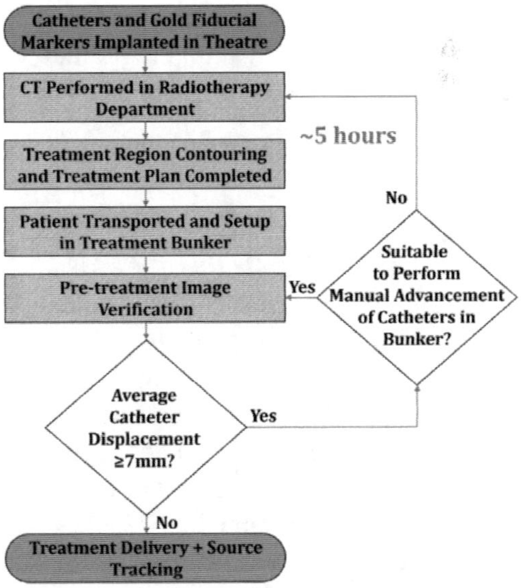

Figure 2. Treatment workflow at AHRO from catheter implantation to treatment delivery.

pre-treatment imaging. The pre-treatment catheter displacement assessment represents only SI movement, which is known to dominate implant movement. It is due to perineal edema shifting the implant inferiorly, so the offset is applied by moving the source dwell positions toward the distal end of the catheters. To assess the dose recovery achieved by the shift, the adapted treatment plan was applied to the anatomical structure information extracted from the re-scan CT and re-plan, to appropriately account for any anatomical changes. Plan metrics were calculated using the in-house developed source tracking dose calculation tool, MaxiCalc (3). MaxiCalc was also used to calculate the dose volume histogram (DVH) indices for the original clinical treatment plan applied to the updated anatomy, to assess the consequence of proceeding with treatment delivery in the absence of any adaption. Each of these was compared to the re-plan metrics to assess the suitability of this first-order adaption strategy. For this investigative study, if the offset exceeded the available space for shifted dwells in the re-scanned catheter, the positions of adapted dwells were extrapolated beyond the length of the catheter.

3. Results

3.1 Catheter Displacement

The mean catheter displacement for the 45 measured catheters was 12.0 mm (range 5.8 mm to 17 mm) in the inferior direction. The average standard deviation for the 9 patients was 1.1 mm (range 0.7-2.1 mm) indicating that implant shift is dominated by a global movement of the entire implant. A displacement exceeding the 7 mm threshold was seen in 44 of the 45 individual catheter measurements, 33 catheters (73.3%) exceeding 10 mm and 9 catheters (20.0%) equal to or exceeding 15 mm.

3.2 DVH Calculation

Figure 3 shows the DVH indices for the adapted plan compared to the re-plan. Also included is the expected outcome if no adaption was implemented i.e. the original plan dwell locations transcribed to the new implant geometry and re-calculated on the updated anatomy and structures. Good agreement can be seen between the adapted plan and that achieved by complete re-planning.

4. Discussion

4.1 Catheter Displacement

The evaluation of catheter displacement is important in the context of HDR brachytherapy for prostate cancer, as it directly impacts treatment efficacy and patient outcomes. In this study, we conducted a retrospective analysis of catheter displacement data obtained from patients that exceeded our clinical treatment threshold for catheter movement as measured using our pre-treatment imaging system. These are the patients for whom a re-scan and re-plan exist for evaluation of the proposed adaption strategy. This cohort featured average catheter tip movements of 12.0 mm while 73.3% suffered inferior displacements greater than 10 mm. This is consistent with studies by other groups, such as Martinez et al. 2001, Holly et al. 2011 and Kawakami et al. 2014 who also document patient cases with average displacements greater than 10 mm (4-6). Pre-treatment imaging to date has shown that 23 of the 47 (48.9%) cases met the re-plan criteria, thus is attractive as an efficient alternative to full re-planning.

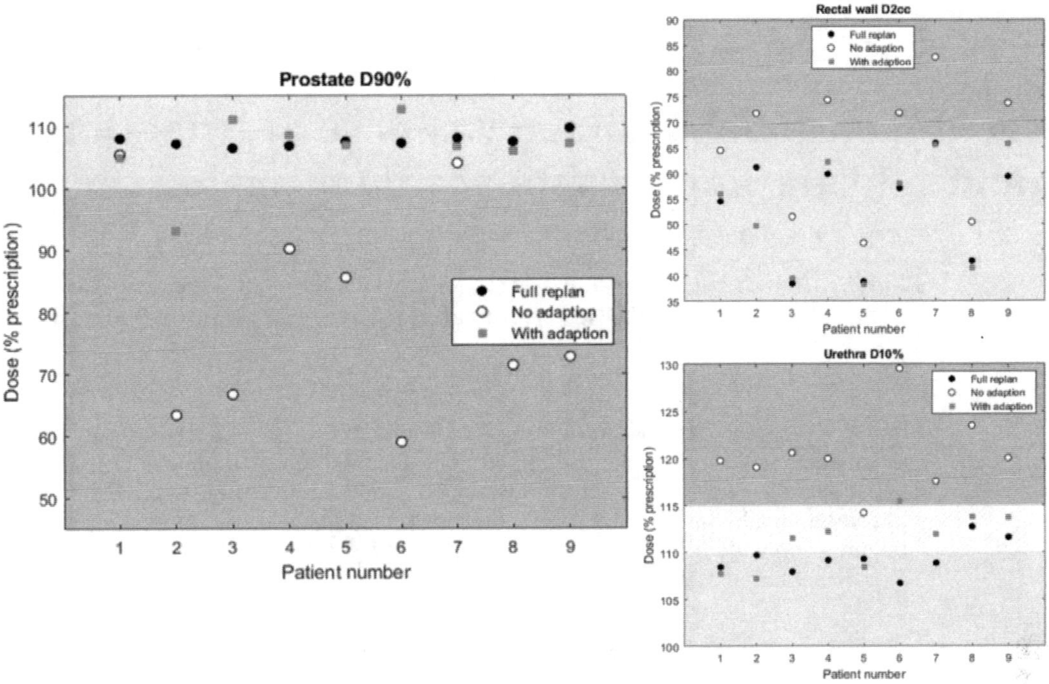

Figure 3. DVH results for individual patients comparing the adapted plan (magenta square) to full re-plan (black circle), and the case where no adaption was implemented (white circle black boarder). Green background indicates within tolerance, red is out of tolerance, yellow is acceptable deviations.

4.2 DVH Calculations

Dosimetric results show that a unilateral shift of the plan dwells is a promising strategy for adapting the original plan to recover planning objectives and is potentially an efficient alternative to replanning. Target coverage planning objectives were met in 8 of the 9 cases studied while OAR sparing was acceptable in 8/9 cases. Our implant assessment and dosimetric evaluation could provide the tools to quickly inform the clinical decision on whether to proceed with this simple adaptation or to re-plan. For practical implementation, consideration will need to be given to the availability of 'head room' space within catheters. Catheter insertion could be increased to provide additional margin.

5. Conclusion

Large implant shifts are frequently observed prior to treatment delivery in HDR prostate brachytherapy. We have demonstrated a first-order adaption strategy that is a viable alternative to time-consuming re-scanning and re-planning and thus improves workflow efficiency.

References

[1] Smith R L, Hanlon M, Panettieri V, Millar J, Matheson B, Haworth A, Franich R 2018 *Brachytherapy* **17** pp 111-121
[2] Smith R L, Haworth A, Panettieri V, Millar J, Franich R 2017 *Physica Medica.* **39** pp 121-31
[3] Hanlon M D, Smith R L, Franich R D 2022 *Physica Medica.* **94** 58-64
[4] Martinez A A, Pataki I, Edmundson G, Sebastian E, Brabbins D, Gustafson G 2001 *Int. J. Radiat. Oncol. Biol. Phys.* **49** pp 61–69
[5] Holly R, Morton G C, Sankreacha R, Law N, Cisecki T, Loblaw D A, Chung H T 2011 *Brachytherapy.* **10** pp 299-305
[6] Simnor T, Li S, Lowe G, Ostler P, Bryant L, Chapman C, Inchley D, Hoskin P J 2009 *Radiother. Oncol.* **93** pp 253-258

Journal of Physics: Conference Series **2799** (2024) 012019 doi:10.1088/1742-6596/2799/1/012019

Three-dimensional dosimetry of dose degradations around gold markers in spot-scanning proton therapy

Lia B Valdetaro[1, 2*], Liliana Stolarczyk[1], Peter S Skyt[1], Stine E Petersen[1], Heidi S Rønde[1], Peter Balling[3,4], Jørgen B B Petersen[5], Ludvig P Muren[1,6]

[1] Danish Center for Particle Therapy, Aarhus University Hospital, Aarhus, Denmark
[2] Department of Oncology, Odense University Hospital, Odense, Denmark
[3] Department of Physics and Astronomy, Aarhus University, Aarhus, Denmark
[4] Interdisciplinary Nanoscience Center, Aarhus University, Aarhus, Denmark
[5] Department of Medical Physics, Aarhus University Hospital, Aarhus, Denmark
[6] Department of Clinical Medicine, Aarhus University, Aarhus, Denmark

*Email: lia.barbosa.valdetaro@rsyd.dk

Abstract. Fiducial markers are necessary for some tumor sites to ensure safe and accurate delivery of radiotherapy including proton therapy (PT). However, the high metal content of the markers may lead to shadowing of the proton beam. The aim of this study was to investigate the dose degradations around three fiducial marker types implanted in different configurations into a 3D dosimeter. The dosimeters were irradiated with a clinically relevant prostate cancer PT plan. Gamma comparisons (2%-2mm) between the control (no marker) and marker dosimeters resulted in a pass rate of 97%, and no significant differences were observed in the isodoses, indicating that markers did not affect dose coverage to the target. However, due to optical artifacts, the 3D dosimeter was unable to resolve the regions within a few millimeters from the markers.

1. Introduction

Fiducial markers are used in radiotherapy for image guidance to ensure accurate patient positioning for tumor sites presenting inter-fractional motion and/or low contrast in planning computer tomography (CT) images [1]. For tumor sites in the pelvis, e.g. prostate cancer, tumors can move up to 2 cm due to bladder and rectum filling [1-3], and high contrast structures, such as bony structures, cannot be used reliably for prostate positioning [2]. Therefore, fiducial markers are implanted inside or nearby the tumor and their position in the daily cone beam CT (CBCT) are compared to the planning CT before treatment. However, fiducial markers may cause shadowing of the proton beam, since they must generally be composed of a high-density material to ensure good visibility on the CBCT [1, 4, 5].

For single spread-out Bragg peak (SOBP) beams, under-dosages as high as 30% have previously been measured with films [1, 6]. However, simulation studies have shown that the shadowing effect is expected to be minimized when multiple fields are used [7]. To the best of our knowledge, this has not been experimentally verified for clinical proton therapy plans.

The aim of this study was therefore to investigate the dose degradation around fiducial markers with respect to target coverage of a patient-specific prostate plan using radiochromic silicone-based 3D dosimeters.

2. Materials and Methods

2.1 Dosimeter fabrication and optical imaging

Two radiochromic silicone-based dosimeters (Ø 10 cm, height 10 cm) were produced; containing (in weight percentages [wt%]), 89.24 wt% silicone elastomer base, 9.0 wt% curing agent (Sylgard 194, Silicone elastomer kit, Dow Corning), 0.26 wt% leucomalachite green (LMG) powder (Sigma-Aldrich) and 1.5 wt% chloroform (Sigma-Aldrich) [8-15]. When irradiated, colorless LMG molecules get converted into malachite green (MG), which absorbs light, making it possible to measure the deposited dose with an optical-CT scanner. The silicone matrix allowed for the insertion of fiducial markers with the same applicators used for patients, so they could be spatially distributed in a clinically realistic way. The dosimeters cured for 48 hours at room temperature (20°C). Optical CT images were acquired two hours before and after irradiation with the Vista 16 (Modus Medical Devices), using 2000 projections over a 360 degrees rotation and 500 ms acquisition time per image [16, 17].

2.2 Gold markers

Two different Gold Anchor marker types (Table 1) were inserted directly into the dosimeter with an applicator, whilst the Best Gold Marker was placed in a cylindrical hole.

Table 1. Description and physical characteristics of the fiducial markers evaluated in this study. The markers are numbered 1-5 for comparison with Figure 2.

Name	Manufacturer	Material	Shape	Length (mm)	Diameter(mm)
Best Gold Marker 785-1	Best Medical International	Gold	Cylindrical	1.5	0.8
Gold Anchor GA200-10	Naslund Medical AB	99.5% gold and 0.5% iron	Folded and straight	10	0.28
Gold Anchor GA200-10B	Naslund Medical AB	99.5% gold and 0.5% iron	Folded and straight	10	0.4

2.3 Treatment planning and irradiation procedure

In this study, we used a test case mimicking a realistic clinical scenario, where prescribed mean doses were 78 Gy to the prostate (with vesicles) and 56 Gy to elective lymph nodes (in 39 fractions, 5 times per week). The plan consisted of four proton fields, with gantry positions of 175, 100, 260 and 185 degrees. The dosimeters were inserted in a Gammex (CIRS) phantom for irradiation. The planned dose was escalated from a single fraction (2 Gy), to 10 Gy, in order to get enough signal in the dosimeters.

2.4 Optical CT data reconstruction

Due to the presence of the gold markers, which are highly reflective, a very careful post treatment of the data was necessary, to minimize optical artifacts in the 3D reconstructed map. First, the raw transmission image files were imported into MATLAB, where a script was used to find the correct match between pre-and post-scans: since the shadows from the gold markers were significantly darker than the rest of the image, a Boolean mask (where the markers were one and the rest, zero) was used to facilitate the image registration. Afterwards, a 3x3 pixel median filter was applied, and the data was imported back into the Vista Software for image reconstruction. Image reconstruction was performed with an OSC-TV algorithm [18], with a 1 mm^3 voxel size.

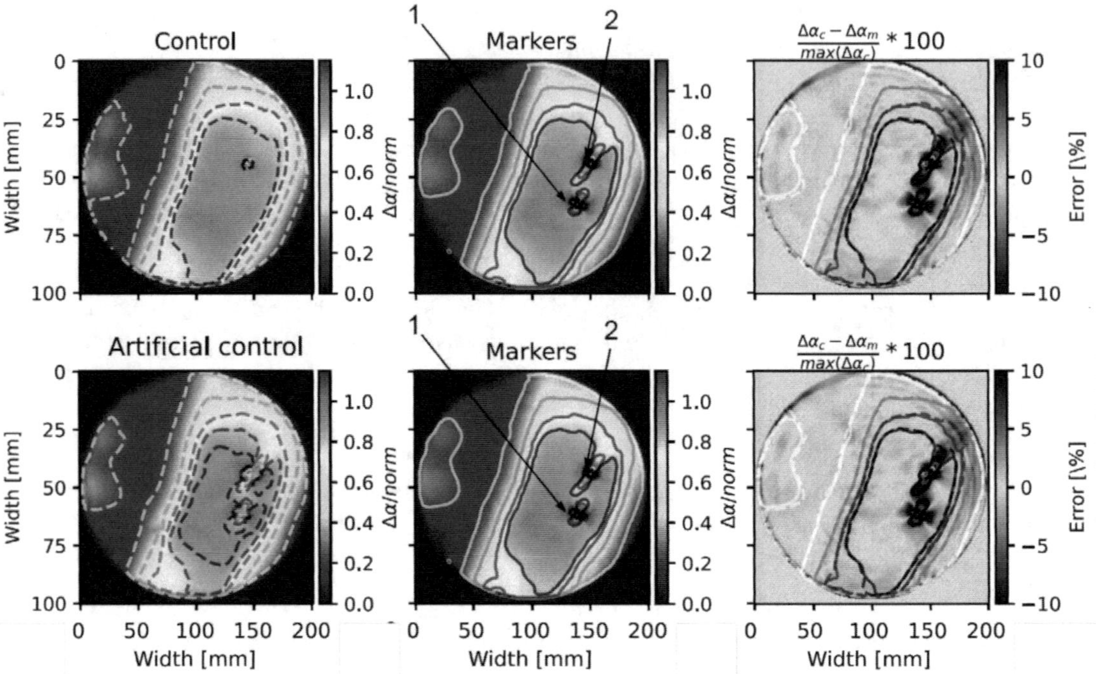

Figure 1: Measurement slice at the marker positions, for the control and marker dosimeters (upper left and center), and *artificial control* (control dosimeter + Boolean mask of the markers) and marker dosimeters (bottom left and center), with the percentage difference between the two (right). Note that the "low dose" region in the control dosimeter (upper left), is the hole for the Best Gold Marker. The marker labels are (1) Best Gold Marker and (2) folded Gold Anchor GA200-10.

2.5 Data Analysis

The reconstructed 3D optical attenuation maps for the "control" and "marker" dosimeters were aligned in MATLAB with an intensity-based image registration algorithm. Local gamma maps of 2%-2mm were calculated (only voxels above 10% of the maximum optical attenuation were included [18]). Although a careful matching between pre- and post-scans removed most reconstruction artifacts, the region a few millimeters from the markers still had what appeared to be visible artifacts in the 3D reconstructed images. Therefore, an "artificial" control image reconstruction, where the *markers were superimposed on the raw image projections of the control* dosimeter (by applying a Boolean mask of the markers to the projection images of the marker dosimeter, and adding the resulting image to the control dosimeter projections), was produced to investigate if it was optical artifacts or a real under-dosage area (Figure 1).

3. Results

Overall, there was very good agreement between the marker and control dosimeters. Gamma pass-rates using 2%-2 mm criteria were 92% between the control and marker dosimeters, and 97% between the artificial control and the marker dosimeters. Optical artifacts in the marker dosimeter were visible 3-5 millimeters from the markers but did not affect the measurement quality in other regions (Figure 1, top). Additionally, a comparison between the control dosimeter with artificial markers indicated that optical artifacts could explain the low and high optical density values around the markers from the image reconstruction algorithm (Figure 1, bottom).

Journal of Physics: Conference Series **2799** (2024) 012019 doi:10.1088/1742-6596/2799/1/012019

4. Discussion

By performing a careful match between pre- and post-scans and applying a mask to the markers, we were able to significantly reduce, but not fully eliminate optical artifacts. This was likely due to residual misalignments, such as the dosimeter being placed slightly differently in the optical CT-scanner in the pre-and post-irradiation scans and/or a small rotational misalignment. Additionally, the cylindrical marker placed in the hole moved a little between the pre- and post-irradiation scans, making it impossible to perfectly align the raw projections.

Due to the significant energy-dependency of the dosimeter used in this study, the measurements were not corrected to dose, with optical attenuation coefficient maps shown instead [11, 15]. This was deemed the best approach, since correcting for the energy dependency would require Monte Carlo simulations, where the markers would have to be precisely positioned in the virtual phantom, which would be very difficult. Therefore, given the additional uncertainties in such a scenario, comparing the measurements of the dosimeter with markers directly to a control seemed to be more relevant. To the best of our knowledge, this was the first experimental study investigating the effect of fiducial markers in a clinical plan for spot-scanning proton therapy. The same markers were previously investigated for mono-energetic proton beams, with marker 1 creating under-dosages of 20-50%, markers 3 and 4 creating under-dosages of 5-15%, and markers 3 and 4 creating under-dosages of 10-20% [6]. Given the high gamma agreement between the control dosimeter with artificial markers and the marker dosimeters, and the fact that the pattern around the markers was nearly perfectly reproduced, it is reasonable to assume that the beam shadowing is minimal. This conclusion would be in line with the findings of Matsuura, *et al.*, who reported in a Monte Carlo study that the use of two or more spot scanning proton fields prevent a reduction in Tumor Control Probability (TCP) for tumors with implanted markers [7].

In conclusion, our results indicated that gold markers did not influence the isodose curves in a realistic scenario with a clinical beam arrangement. However, additional experiments with films will be necessary to confirm the previous assertion.

References

[1] Reidel C A, et al. 2022. *Front. Oncol.* **12** 1–18.
[2] Kotte A N T J, et al. 2007. *Int. J. Radiat. Oncol. Biol. Phys.* **69** 419–25
[3] Kupelian P A, et al. 2006. *Int. J. Radiat. Oncol. Biol. Phys.* **66** 593–6
[4] Zhang M, et al. 2015. *J. Appl. Clin. Med. Phys.* **16** 373–6
[5] Giebeler A, et al, 2009. *J. Appl. Clin. Med. Phys.* **10** 63–70
[6] Stolarczyk L, et al, 2022 PO-1682. *Radiother. Oncol.* **170** S1481–2
[7] Matsuura T, et al. 2012. *Med. Phys.* **39** 5584–91
[8] Høye E M, et al. 2015. *Phys. Med. Biol.* **60** 5557–70
[9] De Deene Y, et al. 2015. *Phys. Med. Biol.* **60** 1543–63
[10] Taasti V T, et al. 2016. *Med. Phys.* **43** 2780–4
[11] Høye E M, et al. 2017. *Phys. Med. Biol.* **62** N73--N89
[12] Jensen M B, et al. 2020. *Phys. Imaging Radiat. Oncol.* **16** 81–4
[13] Jensen M B, et al. 2022. *Acta Oncol.* **61** 264–8
[14] Jensen S V, et al. 2020. *Phys. Imaging Radiat. Oncol.* **16** 134–7
[15] Valdetaro L, et al. 2021. *Phys. Imaging Radiat. Oncol.* **18** 11–8
[16] Doran S J 2010. CT *J. Phys. Conf. Ser.* **250** 424–31
[17] Doran S J 2013. *J. Phys. Conf. Ser.* **444** 012004
[18] Low D A, et al. 1998. *Med. Phys.* **25** 656–61

Extension of a radiation transport model for water-equivalent, portal imaging dose applications

Ivan Kutuzov[1,2], Ryan Rivest[1,2,3], Eric Van Uytven[1,2,3], Boyd McCurdy[1,2,3,4*]

[1] Medical Physics Department, CancerCare Manitoba, Winnipeg, Canada
[2] Department of Physics and Astronomy, University of Manitoba, Winnipeg, Canada
[3] Department of Radiology, University of Manitoba, Winnipeg, Canada
[4] Paul Albrechtsen Research Institute CancerCare Manitoba, Winnipeg, Canada

*E-mail: bmccurdy2@cancercare.mb.ca

Abstract. The amorphous-silicon based design of electronic portal imaging device (a-Si EPID) is commonly available on medical linear accelerators, and thus presents enormous potential as a radiotherapy dosimetry tool. One of the recognized technical challenges with using the device is it's lack of water equivalency when measuring dose. In this work a radiation transport model, previously used to predict dose to the phosphor of an a-Si EPID, was modified to predict dose to a water-equivalent planar detector as well as to a 3D water-tank. Initial testing was performed for a 6MV beam using a variety of simple square fields and clinically relevant intensity modulated fields. Using a 2% criterion, the predicted versus measured image comparisons had pass rates between 95.9-98.2% for the square fields, and 89.2-96.3% for the modulated fields. The predicted 3D dose distribution showed a percentage depth dose agreement within 1% of that measured in a water tank (beyond 5 mm depth). These initial validation results provide confidence that the radiation transport model could be used for water-equivalent dosimetric applications in clinical radiotherapy.

1. Introduction

The amorphous-silicon electronic portal imaging device (a-Si EPID) is commonly available on modern medical linear accelerators and shows many qualities attractive in a dose measurement device [1-3], and thus offers great potential as a radiotherapy dosimeter. However, one challenge is it's lack of water equivalency [4-8]. This can be an impediment to clinical uptake since dose-measuring devices in radiotherapy are typically designed for water-equivalent measurement, and all major dosimetry protocols are water-based (e.g. AAPM TG 51 [9], IAEA 398 [10], AAPM TG 106 [11]). Therefore, the ability to obtain water-equivalent planar dose images or a 3D dose distribution in a water scanning tank using an EPID measurement would enhance it's applicability for dosimetric quality assurance (QA), both patient-specific and linac.

Many have investigated methods to overcome the non-water equivalency of the a-Si EPID. Some have replaced the metal/phosphor screen with a solid water layer [12-14] or plastic scintillator arrays [15-18]. Efforts have also been made developing math models to convert the measured a-Si EPID image to a water-equivalent image [19-24]. These models are typically semi-empirical and have limited robustness in clinical application, restricted to the range of measurements taken (e.g. field size, imager distance, or size of scattering object). Neural and deep convolutional network models have been examined [25, 26] but also have limitations.

In the current work, an existing model-based EPID image prediction algorithm, previously validated for predicting dose-to-phosphor in a-Si EPID detectors, is extended to predict water-equivalent dose images at the EPID plane and also to predict three-dimensional dose distributions in a water phantom. The work is validated here initially using non-transit measurements and a forward dose calculation method.

2. Materials and Methods

2.1 Radiation transport model

A physics-based, radiation transport model [27, 28], previously validated using a-Si EPIDs (aS1000, aS1200; Varian Medical Systems, Palo Alto, USA) was extended. The model steps are:

1) calculate energy fluence produced by the linac, including beam modulating components,
2) if needed, calculate attenuation and scatter effects of any patient in the beam path, and
3) convert incident energy fluence to dose in an arbitrary planar detector at a defined SSD.

For the current work we extended step three, which is described as (equation 6 of [27]):

$$D_{EPID}(x,y) = \sum_i \Phi_{EPID}(x,y,E_i) \otimes k(x,y,E_i) \quad (1)$$

where the dose to the EPID at detector location (x, y) is calculated as the sum of convolutions over i energy bins of the incident fluence estimated at the imaging plane (Φ_{EPID}) with the corresponding dose deposition kernel, k, valid for the detector and photon energy E. The model was tested here in a 'forward' mode (predicting dose to the image detector or to a virtual phantom), but can be used in an iterative reconstruction mode [29], not tested here.

2.2 Extension of radiation transport model

A new library of monoenergetic, 2D dose kernels (0.1 to 6 MeV) was generated for the water-equivalent detector, thus providing k as a function of E in eq (1). These dose kernels are used to convert the fluence incident on the detector to dose deposited in the detection medium. The kernels were generated by Monte Carlo simulation (EGSnrc, DOSXYZnrc) using the known geometry of the ion chamber array (MatriXX, IBA Dosimetry, Schwarzenbruck, Germany). In each simulation, the scoring plane was defined to coincide with the depth and thickness of the ion chambers using a 1×1 mm^2 resolution, which matches the modelled incident fluence. The predicted dose was then down-sampled to match the spatial resolution of the MatriXX detector.

2.3 Validation of extended radiation transport model

The new, water-equivalent image modeling was validated by comparing a series of predicted images against in-air MatriXX measurements without phantom at 100 cm SDD. Ten geometric fields and ten fields with clinically relevant multi-leaf collimator (MLC) configurations for a range of treatment sites were used: prostate intensity modulated radiation therapy (IMRT), head & neck IMRT, and lung volumetric modulated arc therapy (VMAT). Model performance was evaluated by examining the percentage difference and gamma index (using 2%, 2 mm agreement) between the predicted and measured images.

A 3D dose distribution was predicted in a 40×40×40 cm^3 water tank at 100 cm SSD, from a 10×10 cm^2 field using a 6MV flattened beam, based on an EPID measurement. The percentage depth dose (PDD) and beam profiles at the depth of maximum dose (14 mm) were compared with the corresponding measurements made in a similarly configured water tank.

3. Results

The comparison of the water-equivalent predicted and MatriXX-measured planar images using a 2% criterion demonstrated pass rates 95.9%-98.2% for all geometric fields and 89.2-96.3% for modulated fields. Figure 1 shows the comparison for an example prostate IMRT field.
For the example of dose in a 3D water tank, the predicted PDD agreed with the clinical measurements within 1% beyond the first 5 mm of depth in water (Figure 2a) while predicted beam profiles demonstrated 100% pass rate using 2%/2mm criterion (Figure 2b and 2c).

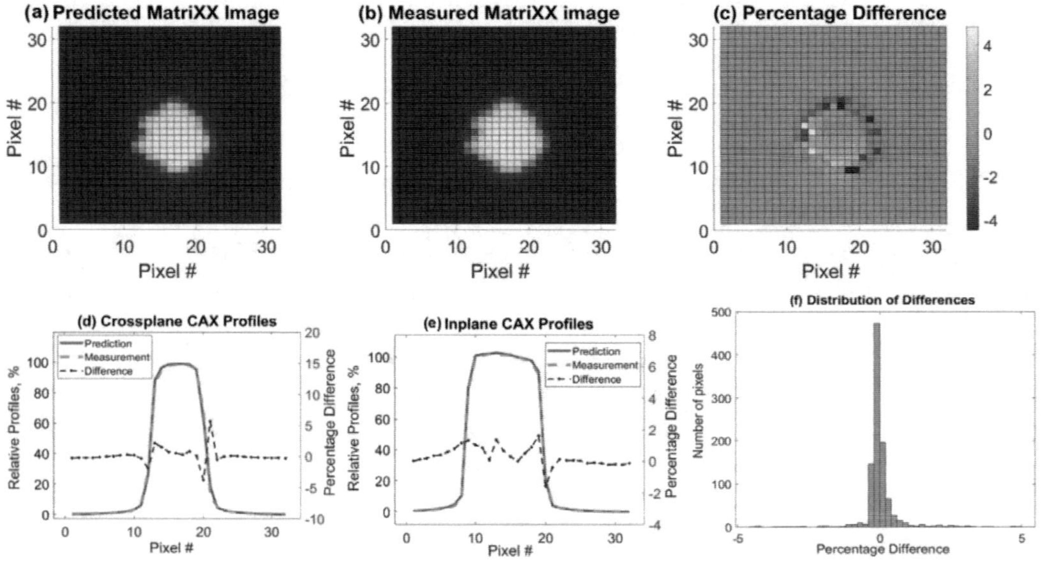

Figure 1. Comparison of an example 2D dose predicted (a) and measured (b) 6 MV prostate IMRT field, percentage difference image (c), cross-plane (d) and in-plane (e) profiles with associated percentage difference, and the histogram distribution of differences (f).

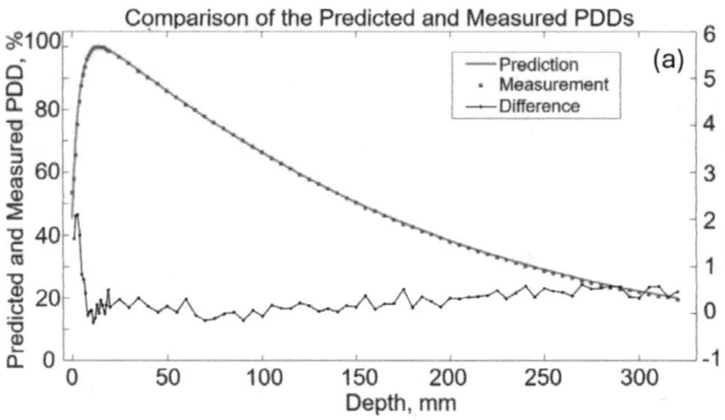

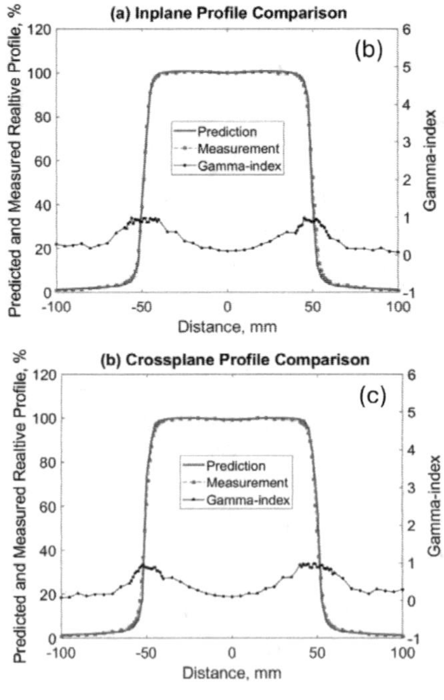

Figure 2. (a) Comparison of the 3D dose predicted and measured percentage depth dose curves for a 6MV, 10x10 cm^2 reference field in a water scanning tank. Gamma-index (2%, 2 mm) comparison of the predicted and measured 3D dose (b) in-plane and (c) cross-plane profiles from a 6MV, 10x10 cm^2 reference field at 14mm depth in water.

4. Discussion and Conclusions

An existing physics-based radiation transport model was extended to predict water-equivalent dose images in a commercial planar detector and volumetric dose distributions in a commercial scanning water tank. A small misalignment in the fields of the predicted versus measured comparison is apparent in the images but is restricted to approximately 5% of central field dose. Due to the low resolution of the MatriXX detector, a gamma-comparison was not used in order to better assess field edge comparisons. However, the pass rates obtained for comparing the predicted water-equivalent dose images using the extended radiation transport model to the water-equivalent dose images measured with a MatriXX detector array are similar to the accuracy obtained for the original model validation using an a-Si EPID [28]. These results are also similar in accuracy to other published water-equivalent conversion models including King et al [21] who obtained gamma pass rates (2%, 2mm) of 92.5-98.7% for IMRT and geometrical fields, as well as Nijsten et al [20] who obtained gamma pass rates (3%, 3mm) of 89.7-97.8%. In summary, a model-based dose image prediction method was extended for water-equivalent imaging systems and was validated with a variety of incident fields. After validating the extended model, we will next implement it to directly convert a-Si EPID images to water-equivalent dose images that can be used in clinical linac QA applications which require water-equivalency, such as flatness and symmetry measurements.

References

[1] Greer P., Popescu C., Med Phys. 2003;30(7):1618-1627.
[2] McCurdy B., Greer P., Med Phys. 2009;36(7):3028-3039.
[3] Chang C-S, Tseng Y-H, Hwang J-M, Shih R, Chuang K-S., Radiation Measurements. 2016;91:9-14.
[4] McCurdy B., Luchka K., Pistorius S., Med Phys. 2001;28(6):911-924.
[5] Kirkby C., Sloboda R., Med Phys. 2005;32(8):2649-2658.
[6] Chatelain C, Vetterli D, Henzen D, et al. Medical Physics. 2013;40(6 Part1).
[7] McCurdy B., Greer P., Bedford J., Electronic Portal Imaging device dosimetry. Clinical 3D Dosimetry in Modern Radiation Therapy, First Ed. 2017:169-198.
[8] Olaciregui-Ruiz I., Beddar S., et al., Phys Imaging Radiat Oncol. 2020;29(15):108-116.
[9] Almond PR, Biggs PJ, Coursey BM, et al. 1999;26(9):1847-1870. doi:10.1118/1.598691
[10] International Atomic Energy Agency. Technical Report Series 398. Absorbed Dose Determination in External Beam Radiotherapy. International Atomic Energy Agency; 2000.
[11] Das IJ, Cheng C, Watts RJ, et al. Med Phys. 2008;35(9):4186-4215.
[12] Sabet M, Menk F., Greer P., Med Phys. 2010;37(4):1459-1467.
[13] Sabet M, Rowshanfarzad P, Vial P, Menk FW, Greer PB., Phys Med Biol. 2012;57(15).
[14] Deshpande S., Blake S., Xing A., Metcalfe P., Med Phys. 2018;45(3):1266-1275.
[15] Guillot M, Beaulieu L, Archambault L, Beddar S, Gingras L., Med Phys. 2011;38(12):6763-6774.
[16] Teymurazyan A, Pang G., Med Phys. 2012;39(3):1518-1529.
[17] Blake S., McNamara A., Vial P., Holloway L., Kuncic Z., Phys Med Biol. 2014;59(22):6827-6840.
[18] Blake S., Cheng Z., McNamara A., Lu M., Vial P., Kuncic Z., Med Phys. 2018;45(5):2154-2168.
[19] Brace OJ, Fuduli I, Alnaghy S, et al., Applied Sciences. 2022;12(2):537.
[20] Nijsten SM, van Elmpt WJ, Jacobs M, et al., Medical Physics. 2007;34(10):3872-3884.
[21] King B., Morf D, Greer P., Med Phys. 2012;39(5):2839-2847.
[22] Zwan B., King B., O'Connor D., Greer P., Med Phys. 2014;41(8):081716.
[23] Miri N, Keller P, Zwan BJ, Greer P., J Appl Clin Med Phys. 2016;17(6):292-304.
[24] Han B, Ding A, Lu M, Xing L., J Appl Clin Med Phys. 2016;18(1):9-17.
[25] Jia M, Wu Y, Yang Y, et al., Med Phys. 2021;48(11):6810-6819.
[26] Bosco LD, Franceries X, Romain B, Smekens F, Husson F, Le Lann M., J Appl Clin Med Phys. 2023;24(6):e13923.
[27] Chytyk-Praznik K., et al., Med Phys. 2013;40(3):031713.
[28] McCowan P., et al., Med Phys. 2015;42(12):6955-6963.
[29] van Uytven E, van Beek T, McCowan PM, et al., Med Phys. 2015;42(12):6945-6954.

Journal of Physics: Conference Series **2799** (2024) 012021 doi:10.1088/1742-6596/2799/1/012021

Development of a new 3-dimensional scintillating detector for patient treatment quality control in pencil beam scanning proton therapy

Anne-Marie Frelin[1*], G. Daviau[1], My Hoang Hoa Bui[1], Cathy Fontbonne[2], Jean-Marc Fontbonne[2], Dorothée Lebhertz[3,2], Erwan Mainguy[1], Cyril Moignier[3,2], Anthony Vela[3,2] and Juliette Thariat[3,2].

[1] Grand accélérateur National d'Ions Lourds (GANIL), CEA/DRF-CNRS/IN2P3, Boulevard Henri Becquerel, 14076 Caen, France
[2] Université de Caen Normandie, ENSICAEN, CNRS/IN2P3, LPC Caen UMR6534, F-14000 Caen, France
[3] Medical Physics Department, CLCC François Baclesse, 14000 Caen, France

*E-mail: anne-marie.frelin@ganil.fr

Abstract. In this work, we developed a three-dimensional (3D) scintillation system, whose fast response and high sensitivity allow the measurement of pencil beam (PB) characteristics (position, energy and intensity in terms of monitor units - MU) for pencil beam scanning proton therapy. The system consists of a $10 \times 10 \times 10$ cm^3 scintillator cube that produces visible light when irradiated with PBs. A high-speed camera records the scintillation distribution and a mirror positioned at 45° to the cube allows visualization of two orthogonal faces of the cube. The measurements demonstrated the ability of our system to measure beam characteristics for intensities as low as $2 \cdot 10^{-3}$ MU in a single irradiation. Standard deviations of less than 300 μm were found for the X and Y PB positions, of approximately 150 keV for energy and of less than $5 \cdot 10^{-3}$ MU for intensity. These measurements were then successfully used to verify the compliance of the PBs delivery with the treatment plan, thus making our system a fast and efficient verification tool.

1. Introduction

Pencil beam scanning Proton therapy has a number of advantages, in particular with regard to the dose deposition profile of charged particles. It consists of scanning the tumour volume with a finite number of pencil beams (PBs) whose energies and positions allow a uniform dose distribution. However, with the IBA Proteus®ONE system [1], the accuracy of the delivered pencil beams in terms of delivered monitor units (MU) requires that the PBs be divided into multiple pulses of 7 μs delivered at a frequency of 1 kHz. These delivery constraints make it difficult to perform patient-specific quality control (QC) based on the verification of the delivered PB characteristics (position, energy, intensity in terms of MU) on a PB-by-PB basis. To address this issue, scintillators are very promising due to their advantageous properties: spatial resolution, real-time measurement, shape versatility [2-7]. Developments have been made in the field of proton therapy with plane scintillators [8], and the Lynx system is commercially available for 2D relative dosimetry and daily quality assurance (IBA dosimetry, Schwarzenbruck, Germany).

Journal of Physics: Conference Series **2799** (2024) 012021 doi:10.1088/1742-6596/2799/1/012021

Developments have also been made with liquid scintillators to measure the range and the position of proton spots [9,10]. Thanks to fast cameras, recent studies have brought scintillation dosimetry into the field of real-time PB characterization, in particular to perform range and dose rate measurements [11,12]. In this work, we developed a three-dimensional (3D) scintillation system, called SCICOPRO (for SCIntillation for treatment COntrol in PROton therapy), whose fast response and high sensitivity allow treatment QC by measuring, in a single irradiation, the position, energy and intensity of the delivered beams on a PB-by-PB and even pulse-by-pulse basis.

2. Materials and methods

The system developed in this work, shown in Figure 1, consists of a scintillator cube of $10 \times 10 \times 10$ cm^3, that produces visible light when irradiated with PBs. A high speed camera records the scintillation distribution and a mirror placed at 45° to the cube allows to visualize two orthogonal faces of the cube, and thus two orthogonal projections of the scintillation distribution, on a pulse-by-pulse basis. The synchronization of the camera with the pulse delivery ensures that each pulse is captured without any beam loss.

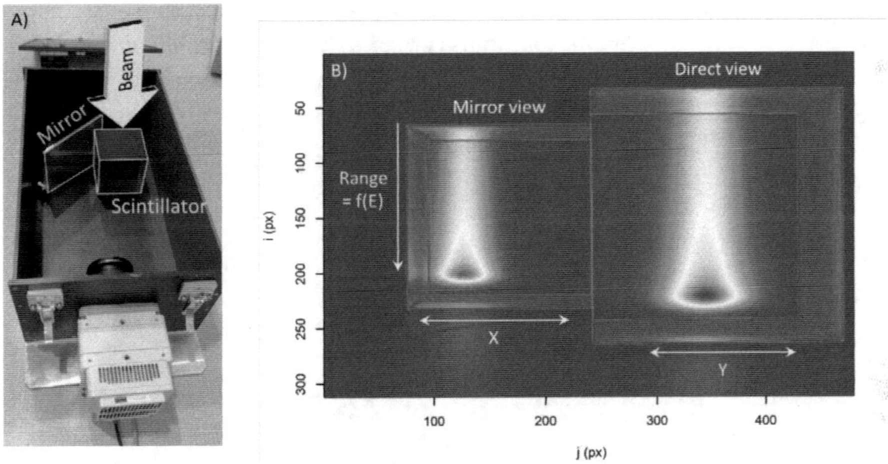

Figure 1. A) SCICOPRO setup consisting of a $10 \times 10 \times 10$ cm^3 BC-412 plastic scintillator, a mirror oriented at 45° to the cube and a fast camera. B) The images recorded by the camera consist of a direct projection and a mirror-reflected projection of the scintillation distribution produced by each pulse in the scintillator.

Pulse images were individually corrected for background signal and hot pixels (a parasitic signal caused by scattered radiation hitting the camera sensor directly). In this application, the PBs were delivered far enough away from the sides of the scintillator to avoid any bias caused by optical reflections at the edge of the scintillator. Analysis of these images provided the position and intensity of the Bragg peak in both projections. The SCICOPRO measurements were calibrated against the Proteus®ONE ionization chamber measurements recorded in log-files because, to our knowledge, this is the only detection system capable of providing all pulse characteristics for cross-calibration.

In a first step, SCICOPRO performance was evaluated on 729 PBs at fixed intensities of 0.02, 0.1 and 1 MU. QC of patient treatment plans was then performed with SCICOPRO for 7 irradiation plans: one delivering a uniform dose of 0.8 Gy in a 5 cm diameter sphere and 6 patient treatment plans. The sphere irradiation plan was delivered, with a nominal positioning, but also with 1°

rotations of the treatment table to evaluate the ability of SCICOPRO to detect potential treatment delivery problems, in this case a positioning error. The PB characteristics were calculated from the pulse characteristics, where the position and energy are the average of the intensity-weighted pulse positions and energy, and the PB intensity is the sum of the pulse intensities. Finally, the measured characteristics of the delivered PBs were compared with the characteristics of the planned PBs. A global agreement factor of the treatment delivery was defined as the proportion of PBs whose characteristics matched the planned ones, with a tolerance of 1 mm for position and 5 % for intensity.

3. Results

The intensity of all PBs of the 7 treatment plans evaluated in this study ranged from 0.02 to 4.1 MU. However, splitting these PBs into several pulses resulted in the delivery of pulses of much lower intensity, all between approximately $2 \cdot 10^{-3}$ to $150 \cdot 10^{-3}$ MU, which drastically reduced the measurement dynamics. It should be noted that SCICOPRO demonstrated its ability to measure pulse characteristics at these very low intensities. Standard deviations of less than 400 µm in X and Y position, of about 180 keV in energy and of less than $0.65 \cdot 10^{-3}$ MU in intensity were measured on the pulses. For PB characteristics, calculated from pulse characteristics, this corresponds to standard deviations of less than 300 µm and 150 keV, regardless of PB intensity. The standard deviation of the measured intensity, on the other hand, was found to depend on the planned intensity and was of about $1 \cdot 10^{-3}$ MU for 0.02 and 0.1 MU and about $5 \cdot 10^{-3}$ MU for 1 MU. This is due to the higher number of pulses required to achieve 1 MU than 0.02 and 0.1 MU.

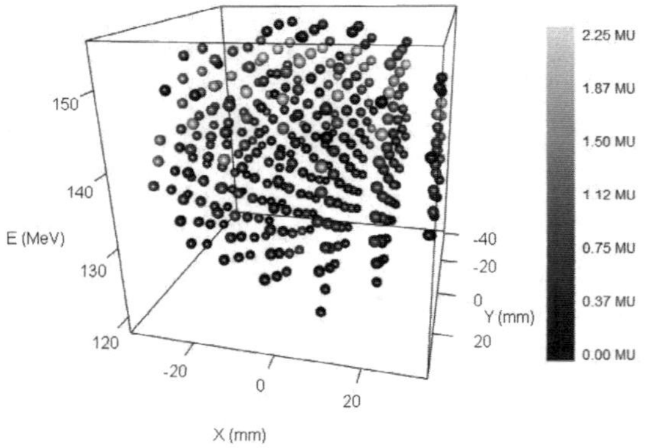

Figure 2. PBs characteristics measured with SCICOPRO for the irradiation plan producing a uniform dose distribution of 0.8 Gy in a 5 cm diameter sphere. Each PB is represented spatially by its X and Y positions and energy (vertical axis). The PB intensity is represented by the colour scale. For clarity, only the measured characteristics are shown here as they closely match the planned characteristics.

An example of PB characteristics reconstructed from pulse characteristics measured with SCICOPRO is shown in Figure 2. This representation, which spatially represents the position and the energy, and in the colour scale the intensity has the interest of displaying the information of all PBs of a treatment, complementary to the 3D dose distribution, which may hide the delivery details. For all the treatment plans tested, SCICOPRO made it easy to verify that the delivered PBs matched the treatment plan. More than 99 % of the PBs were within a tolerance of 1 mm for position and 5 % for intensity. When a 1° rotation of the treatment table was introduced, this proportion fell to 86 % or less, confirming the ability to detect positioning errors.

4. Discussion

The ability to measure proton spot characteristics has been demonstrated previously with liquid scintillators, but at a limited rate of 85 frames per second [9,10]. On the other hand, recent advances in high-speed cameras have been used to perform dose rate and proton range measurements [11,12]. With our system, we demonstrated the feasibility of verifying the characteristics of all PBs of a treatment in a single irradiation, even at very low intensities. The treatment verification performed in this study provides complementary verifications to 2D or 3D dose distributions integrated over the entire treatment, as it can highlight some problems that may be more difficult to identify in integrated measurements.

5. Conclusion

We developed a new scintillating system capable of measuring pulse characteristics at very low intensity (down to $2{\cdot}10^{-3}$ MU) and high frequency (1 kHz), overcoming the delivery constraints of the Proteus®ONE system. It is a very efficient QC tool thanks to its geometry, which allows all PB characteristics of a treatment to be determined in a single measurement.

Acknowledgment

This project has received financial support from the Region Normandie through the "RIN recherche" program 2018 – CPIER – 18E01804/18P03186.

References

[1] Marchand B et al 2000 IBA proton pencil beam scanning: an innovative solution for cancer treatment *Proceedings of EPAC 2000, Vienna, Austria.*
[2] Beddar A S et al 1992 Water-equivalent plastic scintillation detectors for high-energy beam dosimetry: I. Physical characteristics and theoretical consideration *Phys Med Biol.* **37** 1883-900.
[3] Beddar A S et al 1992 Water-equivalent plastic scintillation detectors for high-energy beam dosimetry: II. Properties and measurements *Phys Med Biol* **37** 1901-13
[4] Beddar A S et Beaulieu L 2016 Scintillation dosimetry CRC Press, 2016
[5] Alexander D A et al 2020 Scintillation imaging as a high-resolution, remote, versatile 2D detection system for MR-linac quality assurance *Med. Phys.* **47** 3861-3869.
[6] Frelin AM et al 2008 The DosiMap, a new 2D scintillating dosimeter for IMRT quality assurance: characterization of two Cerenkov discrimination methods *Med Phys.* **35** 1651-62.
[7] Frelin AM et al 2005 Spectral discrimination of Cerenkov radiation in scintillating dosimeters *Med Phys.* **32** 3000-6.
[8] Boon S N et al 1998 Fast 2D phantom dosimetry for scanning proton beams *Med Phys.* **25** 464-75.
[9] Darne CD et al 2017 Development of a liquid scintillator-based 3D detector for range measurements of spot-scanned proton beams Phys.: Conf. Ser. 847 012020.
[10] Archambault L et al 2012 Verification of proton range, position, and intensity in IMPT with a 3D liquid scintillator detector system *Med Phys* **39**(3) 1239-1246.
[11] Goddu S M at al 2022 Synchronized high-speed scintillation imaging of proton beams, generated by a gantry-mounted synchrocyclotron, on a pulse-by-pulse basis *Med Phys.* **49** 6209-6220.
[12] Rahman M et al 2020 Characterization of a new scintillation imaging system for proton pencil beam dose rate measurements *Phys Med Biol.* **65**(16).

AUTHOR INDEX